AF581232

A SURVEY OF

Test Methods in Fluid Filtration

GULF PUBLISHING COMPANY
HOUSTON, LONDON, PARIS, ZURICH, TOKYO

A SURVEY OF Test Methods in Fluid Filtration

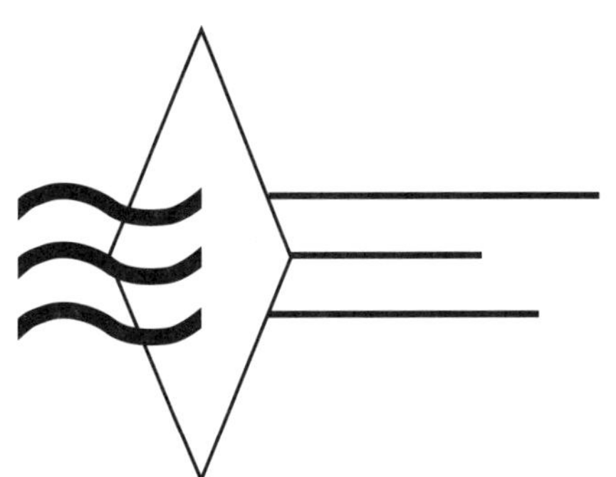

PETER R. JOHNSTON

A Survey of Test Methods in Fluid Filtration

Printed in the United States of America.

Gulf Publishing Company
Book Division
P.O. Box 2608 ■ Houston, Texas 77252-2608

10 9 8 7 6 5 4 3 2 1

Library of Congress Cataloging-in-Publication Data
Johnston, Peter R.
A survey of test methods in fluid filtration/Peter R. Johnston.
p. cm.
Includes bibliographical references and index.
ISBN 0-88415-816-0
1. Filters and filtration. I. Title.
QD63.F5J64 1995
542′.6—dc20 94-37795
CIP

Printed on Acid-Free (∞) Paper

Contents

Preface

About three years ago the Education Committee of the American Filtration and Separation Society (AFS) asked members to send in descriptions of test methods for filtration. It was the aim of that committee to summarize those test methods and thereby provide a reference which would address questions concerning that specific separation process.

This reference, it was hoped, would guide an investigator searching for the appropriate test methods during his or her quest for the best filter medium and for the best filter aid for whatever particles have to be separated from whatever fluid. This reference would also provide a general understanding of filtration. Indeed, it was also hoped, this reference might even help resolve differences of opinion among filter users, filter manufacturers, and those who write standard test methods.

The first draft of this reference was reviewed by more than a dozen members of the AFS, representing members of academia, manufacturers of filter media, and users of filter media. Accordingly, the present narrative represents my best effort at stating what the authors, listed in the bibliography, say about test methods, and, indeed, about filtration in general.

Not all the listed papers are cited in the text. We hope authors who are not cited will not be discouraged from bringing test methods, and results, to our attention in the future.

This book is aimed at a wide range of readers, who can be placed into four categories:

1. the laboratory investigator or plant engineer (with perhaps more on his or her mind than a filtration step) who simply wants some bottom line directions and answers
2. the manufacturer of filter media, who wants more details to help him or her supply the best filter medium
3. the theoretician, who wants to see all the reasoning behind the how and the why of a test method
4. those groups that write standard test methods

In directing this narrative to these different kinds of readers, I use two types of subheads in each chapter. One is *INSTRUCTIONS,* which addresses *what to do.* The other is *BACKGROUND* which explains *why.*

These subheads point the reader to the separate subjects of actions and theoretical reasoning. Of course, the dividing line between these two subjects is not always simple, and the reader will see some cases where the subjects have been blended. Further, in some chapters one or the other subhead has been excluded, simply because it does not apply to the material being presented.

The many reviewers and I have made an honest effort to extract all information from the works cited in the bibliography and to present and discuss this information in this book. However, because this book was neither reviewed by nor subjected to balloting by the entire membership of the AFS, it must not be construed as a statement by the AFS as a whole.

We ask readers who see errors in this book, or omissions, or "clear-as-mud" explanations, to let us know by writing the publisher. Your letters will help us write a second edition as new developments unfold. Indeed, in this book we point out some specific tests and investigations for which we have not seen results—and we would like to see them.

Peter R. Johnston

Introduction

While the hardware of a filter system, including the means of holding the filter medium in place, is certainly important in the design and operation of a system, the item of direct interest is the filter medium itself. The first, and obvious, questions concerning a filter medium are:

- Will it stand up to the fluid and temperature of concern?
- Does it have the required mechanical strength?
- Does it contain materials that wash out into the filtrate?

These questions are not addressed in this book, since they are so specific to the many different filtration problems that can occur. Readers are reminded, however, to answer these questions.

What this book does address are the method of characterizing filter media, the different kinds of filtration tests, and the various kinds of operations.

Because a given filter medium will perform differently under different conditions, there can be no single, universal-standard filtration test. The different conditions include the properties of the fluid, such as its very nature, viscosity, temperature, and velocity, along with the nature of the solids that the medium is attempting to separate from the fluid. But we can assign a rating or category to a filter medium by addressing five points:

1. the material(s) of construction
2. the thickness
3. the porosity (the ratio of void volume to bulk volume)
4. the viscous permeability (deduced from the ratio of fluid flow rate to driving pressure)
5. whether porosity changes with depth.

As stated in the preface, this book addresses different kinds of readers, that is, users of filter media, manufacturers of filter media, and writers of standard test methods. And, as just stated, those readers must consider specific subjects not addressed in this book, namely, the cleanliness of a filter medium, its mechanical strength, and its compatibility with the fluid of concern.

The reader, having then addressed those specific subjects, and having therefore selected certain media for subsequent considerations, will then be able to look to the information in this book for further help in selecting, or specifying, the best medium for the filtration job at hand.

This book presents the subjects of concern in the following order:

We start with a discussion of the physiscs of flow of particle-free fluid through filter media and an explanation of the basic term *permeability.*

We address permeability as a function of packing density for a porous material built from a random array of fibers or granules.

This book addresses concepts of pore sizes as well as pore-size distributions and explains the mechanical and mathematical models describing such distributions along with fluid-intrusion techniques for measuring such distributions. Three different distributions are described: number, volume, and fluid flow—each with a different average-sized pore.

The various meanings of particle size are addressed as well as the different views of particle-size distributions.

Writers who have measured and reported filtration efficiency have used a wide variety of terms (beside filtration efficiency) to address the comparison of the particle-size distribution in the feed stream to that in the filtrate. We here attempt to make sense of that babel.

In discussing the capacity of a filter medium, meaning the mass of particles it can fed before it suffers a significant loss in permeability or the fact that it collects a cake too deep for the apparatus in which it is held, we address the three schemes of filtration, namely, constant fluid-flow rate, constant fluid-driving pressure, and variable flow and pressure. Where the collected particles are to be recovered, we explain the methods of deducing the resistance of the filter cake including some cases where the cake contains a filter aid.

Where three (instead of two) streams are involved, cross-flow filtration is discussed. We address microfiltration, ultrafiltration and reverse osmosis, along with the special vocabulary of those processes.

We end with a short review of separating oil from water.

A SURVEY OF

Test Methods in Fluid Filtration

1
Liquid Flow Through Filter Media
The Meaning of Permeability
Concepts of Pore Size

1.1. INTRODUCTION

INSTRUCTIONS

Probably the first thing one wants to know about a filter medium meant to clarify a liquid is the relationship between the liquid flow rate and the driving pressure.

When the filter medium is a flat sheet, one of the simplest setups to take these measurements involves holding a two-inch diameter disk of the medium under a column of clear water and measuring the flow rate. One filter manufacturer calls this a *rapidity test,* since it measures how fast the water flows under a constant head of two and (or) ten inches of water.

When the filter medium is larger or in the form of a cartridge, a pump is usually involved along with (1) a suitable housing for the medium, (2) a flowmeter, and (3) pressure gauges on either side of the housing. Frequently, the exit stream merely discharges at atmospheric pressure (zero gauge pressure) so that the pressure drop across the housing (the driving pressure) corresponds to the gauge reading on the feed-stream side. But make sure you measure the pressure drop across the cartridge alone.

To avoid including the pressure drop across the housing in this measurement, begin by flowing the liquid through the empty housing at different flow rates, and recording the various pressure drops. Then, with the cartridge in place, and for a given flow rate, subtract the housing pressure from the pressure read with the cartridge, to obtain the pressure drop across the cartridge alone.

Having thus measured liquid flow rate as a function of driving pressure, at three or more driving pressures, plot the data pairs on log/log paper. It doesn't matter what units of measurements you employ; go ahead and make the plot. If these points define a straight line with a slope of 1.0, as illustrated by the left portion of Figure 1.1, then your measurements are over. You have made the measurements in that range where the ratio of flow rate to driving pressure is constant. (Or, instead of making a plot, look at the ratios of flow rate/pressure to see that the ratios are equal.)

If your slope is less than 1.0, as illustrated by the right portion of Figure 1.1, you have imposed more driving pressure, and obtained more flow rate, then the filter medium is meant to handle. Take more measurements at lower driving pressures.

When the filter medium contains very small pores, make sure the liquid is clear enough so that the medium does not plug with solids during the tests; e.g., start by imposing a low driving pressure. After a subsequent test at high pressure, return to the low pressure, to see if the original flowrate is repeated.

BACKGROUND

On making a plot, on log/log paper, of the velocity, u, with which the liquid approaches, or leaves, a face, versus the pressure drop, ΔP, and employing a broad range of driving pressures, you will see the kind of plot shown in Figure 1.1, consisting of two lines, joined with a curve. Recall, velocity, $u = Q/A$ = [volumetric flowrate]/[area of the medium]. The plot can be described (ASTM F 902) by

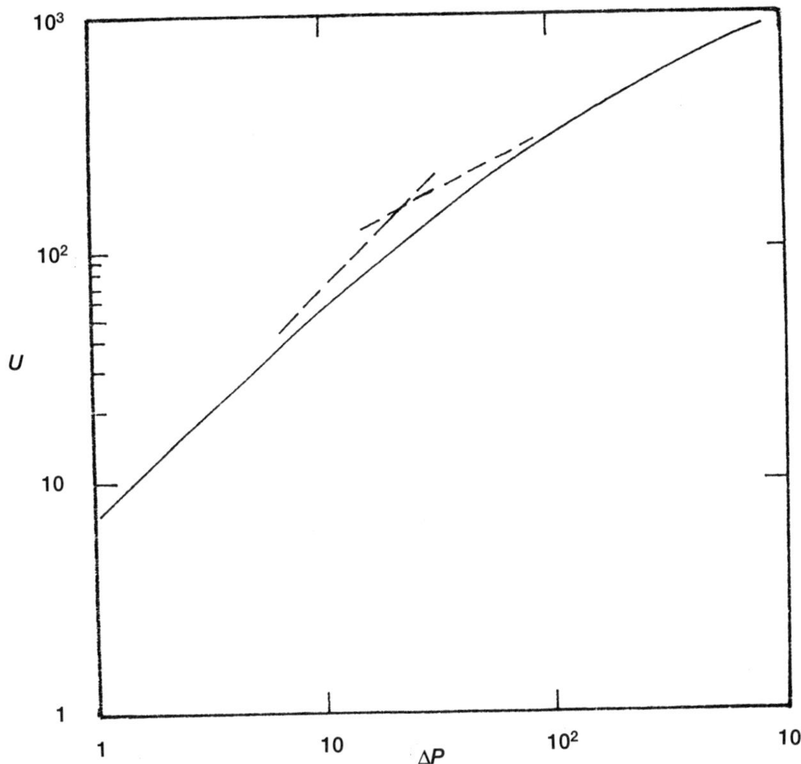

Figure 1.1. Example of liquid-permeability measurements. A plot, on log/log paper, of the velocity, *u*, with which a liquid approaches the face of a filter medium, versus the differential pressure across the two faces, ΔP. The units of measurement are arbitrary in this example, but we see the slopes described by Equation 1.1. How fast the slope falls, from 1.0 to 0.5, with increased driving pressure, is described by Equation 1.2. In filtration, one ordinarily employs viscous flow, that region where the slope is 1.0

$$\frac{\Delta P}{z} = \alpha \eta u + \beta \rho u^2 \tag{1.1}$$

where: ΔP = the differential pressure, N/m^2
z = thickness of the medium, m
α = viscous-term coefficient, m^{-2}

η = absolute viscosity of the liquid, N • s/m^2
u = approach velocity of the liquid, m/s
β = inertia-term coefficient, m^{-1}
ρ = density of the liquid, kg/m^3•g_c (g_c= 1 kg • m/N•s)

While Equation 1.1 suggests plotting ΔP on the vertical axis, we plot it, here, on the horizontal axis, as a reference for discussions coming later.

The line of Figure 1.1 changes slope from 1.0 to 0.5 at the rate defined by (Green & Duwez 1951)

$$f = 2/Re + 2 \tag{1.2}$$

where: f = friction factor $= \dfrac{2\Delta P/z}{\beta\rho u^2}$

= (total forces)/(half inertia force) and

$$Re = \text{Reynolds number} = \frac{\beta\rho u^2}{\alpha\eta u} = \frac{\beta\rho u}{\alpha\eta}$$

= (inertia force)/(viscous force).

1.2 SEPARATING VISCOUS FLOW FROM INERTIA FLOW

BACKGROUND

Where, in Figure 1.1, we see the slope of 1.0, we see the symptom of viscous (laminar) flow. Viscous drag is the predominate resistance to flow. At higher velocities, where the slope is 0.5, we see the symptom of inertia flow, in which the liquid is constantly asked to change direction, and in which viscous drag constitutes a relatively small resistance to flow.

Where the slope is changing, and, indeed, when much of the resistance to flow lies with inertia, the flow of liquid is still laminar (Rosenstein et al. 1980). It is not turbulent as some writers have surmised. Indeed, the Reynolds number (see Equation 1.2) is still low. (Recall, turbulent flow in a pipe is associated with Reynolds numbers over 3,000.) That is, when a single fiber is held in a stream, we see eddy currents around it; this is not so in a fibrous mat.

By *laminar flow* is meant the kind of fluid-velocity profile seen in a pipe: the greatest velocity is in the center, and the velocity falls, approaching zero, toward the wall.

Given the data of Figure 1.1, we can deduce the values of α and β in Equation 1.1 by making a plot, (not shown here) on linear coordinates, of $\Delta P/zu$ versus u, to define the line

$$\frac{\Delta P}{zu} = \alpha\eta + \beta\rho u \tag{1.3}$$

where the intercept is $\alpha\eta$, and the slope is $\beta\rho$.

The ratio β/α is not the same from one medium to another. And, while we can assign a meaning to α (see Section 1.3) we cannot do so with β.

But, assigning a meaning to β (other than calling it the inertia-term coefficient) is not important because filtration is usually done at low velocities, where, at the start (before the medium starts to plug with collected solids) pressure drop and velocity are direct functions of one another; the β term is negligible. Thus, an investigator may not be interested in values of β. He or she only wants to consider the viscous-flow region.

Sometimes, as in the case of pleated-paper filter cartridges, we don't know the area of the filter medium. In that case we plot volumetric flow ate (rather then approach velocity) versus driving pressure, to find the region wherein the slope is 1.0.

1.3 THE MEANING OF PERMEABILITY

INSTRUCTIONS

The data presented above regarding the ratios of flow rate to driving pressure will obviously be different for a liquid of another viscosity. Thus, to characterize a filter medium for any liquid, we deduce the viscous *permeability* of the medium. It is important to understand this term, since

- it is a basic characteristic of a filter medium, and

- we use it in deducing the *flow–average pore size.*

With a flat-sheet medium, and when the flow is directly proportional to the driving force (the slope is 1.0 in the Figure-1.1 kind of plot), we deduce viscous permeability, B (units: m^2) as follows:

$$B = \frac{1}{\alpha} = \frac{u\eta z}{\Delta P} = \frac{Q\eta z}{A\Delta P} \tag{1.4}$$

where: Q = volumetric flow rate of liquid, m^3/s,
A = area of the filter medium, m^2,
Other terms are defined below Equation 1.1.

When the filter medium is a thick-walled tube, or candle, of outside radius r_1, inside radius r_2, and length L, and the liquid flows from the outside wall to the inside wall, we calculate permeability as follows:

$$B = \frac{Q \ln (r_1/r_2)}{\Delta P 2\pi L} \tag{1.5}$$

where *ln* indicates natural logarithm.

However, be careful in taking permeability measurements with a long (e.g., 20–40-inch) cartridges. The flows through the walls are not uniform. The measurements are distorted by the pressure drop along the length of the inner diameter and the hydrostatic head (when the cartridge is vertical). We should think that someone would have already addressed this subject; but we have yet to see even a theoretical study.

Previous writers have expressed permeability in darcies. The conversion to SI units is: 1 darcy = $0.987 \cdot 10^{-12} m^2$, since, by definition, B, in Equation 1.4, equals one darcy when:

Q = 1cc/s = $10^{-6} m^3/s$
A = 1 cm^2 = 10^{-4}m
η = 1 centipoise = $10^{-3} N \cdot s/m^2$
z = 1 cm = 10^{-2}m
ΔP = 1 atmosphere = $1.103 \cdot 10^5 N/m^2$

1.4 CONCEPTS OF PORE SIZE

BACKGROUND

Before describing how to deduce the flow-average pore size, this section addresses what it is we are measuring.

The key to a direct understanding of pore size lies in examining that singular product the track-etched membrane. Viewing that membrane with a microscope, we see that it contains neat, circular tunnels going straight through it. Thus, we can describe a pore as having a certain diameter and a certain length. And, for the most part, all the pores we view have the same diameter and length, the latter being equal to the thickness of the membrane, usually 10 μm.

Occasionally we see where two or more holes are crowded together to make one larger opening. We can also see that about 10% of the membrane surface constitutes the cross-sectional area of the pores. Described in three-dimensional sense, 10% of the bulk volume of the membrane is void space; i.e., the *porosity* of the membrane, ε, is 0.10.

1.5 COMPLICATED PORE GEOMETRY

On examining a microporous membrane manufactured by the solvent-cast method, we see the porosity to be about 0.75. In addressing pore diameter, the only way we can come up with some linear dimension is to choose between either

1. the ratio of cross-sectional area to perimeter, or
2. the square root of the cross-sectional area.

We also notice that not all pores are the same size. Indeed, we see a relatively broad distribution of pore sizes. That is, the ratio of the size of the "largest" pore to the size of the "smallest" is very much greater than the ratio we see in the track-etched membrane.

Further, microporous membranes are about fifteen times as thick as a track-etched membrane. So while we can easily speak of the average pore size in the track-etched membrane as being the size of one of the many, same-sized pores, we realize that with the microporous membrane we must consider the meaning of the *average* pore size.

The same reasoning holds for fibrous mats, which also have porosities near 0.75, with greater porosities for the filtration of gases. But there's more. As we consider how a fluid flowing through these additional media defines different paths, or tunnels, we realize that within each tunnel the diameter varies along the length. And different tunnels have different lengths. (The average length of the tunnels is greater than the thickness of the medium.) Thus, to speak of an average pore size we must consider the average of many distributions.

When the filter medium is a bed of granules, we have the same problem in defining pore size as we have with the microporous membranes and the fibrous mats. Further, the problem of deducing an average-sized pore grows more complex when the porosity of the bed changes with depth. Indeed, ordinary filter paper has a greater packing density of fibers on one face. The machine-forming "wire" side is denser than the other face. And thick-walled, tubular cartridges often have differences in densities across the thicknesses of those walls.

1.6 PORE SIZE AND POROSITY

On building a filter medium by packing given-sized granules, or fibers, closer together, thus reducing the porosity of the filter medium, we decrease the diameter of the average-sized pore (see Chapter 3). In addition, while we can build fibrous beds with a broad range of porosities, we can't do so with granules. If the granules are too far apart, they can't stick together. Filter media built of granules usually have porosities in the range of 0.2 to 0.4.

Some writers, instead of using the term *porosity* use *solidity*, or *solidarity*, which refers to the ratio of the volume of solids to the bulk volume of the medium. Some writers employ *solidosity*, apparently to rhyme with *porosity*.

1.7 THE "LARGEST" AND THE "SMALLEST" PORES

In tests described later we have the tools to look at the distribution of pore diameters. By way of explaining this concept, consider two different pore tunnels where in each the diameter varies along the length. In the case where both tunnels have the same average diameter, we consider that both tunnels have the same diameter. Thus, when we speak of a pore-size distribution we mean a distribution of individual-average pore diameters.

When filter media are built of a random array of "building blocks," the pore-size distribution somewhat follows a log-normal distribution. And for us to speak of the "largest" pore, as many writers do, we must consider the range of the distribution we *want* to embrace—or, indeed, that we are *able* to embrace with the test setup at hand (see Chapter 5).

1.8 DIFFERENT KINDS OF AVERAGE PORE DIAMETERS

In considering a pore-size distribution, and the corresponding *average* pore size, we must decide which of the three following distributions to address:

- In the *number* distribution we consider a list of the counts of pores of different diameters. We measure such a distribution by examining a face of the filter medium.
- In the *volume* distribution the cross-sectional opening is said to have "unit" depth, in that the volume of a pore is proportional to the square of the diameter, d^2. We measure the volume distribution with a mercury-intrusion test or a liquid-drainage test (see Chapter 5).
- As we explain in Section 1.9, when a fluid in viscous flow, under a given driving force, confronts a pore, the volumetric flow rate of that fluid through the pore is proportional to the square of the cross-sectional area, d^4. We try to measure the distribution of *fluid flow* via an extended bubble-point test (see Chapter 5).

Thus, the *flow-average* pore diameter is larger than the *volume-average*, which, in turn, is larger than the *number-average*. Yet, some writers say the volume-distribution of pore diameters is the same as the fluid-flow distribution. And, many writers report "pore size" without explaining what they mean (radius or diameter) or how they measured it.

1.9 DEDUCING THE FLOW-AVERAGE PORE DIAMETER FROM PERMEABILITY AND POROSITY

INSTRUCTIONS

Recall the Hagen-Poiseuille Law for viscous flow of fluid through a tube: For a fluid of viscosity η, the average velocity, u, of that fluid through a tube, is related to the tube diameter, d, under a pressure drop of ΔP, along a length, z, as

$$\frac{d^2}{32} = \frac{u \eta z}{\Delta P} \tag{1.6}$$

We now employ Equation 1.6 to reach a meaning of the average pore diameter defined by a liquid flowing through a filter medium. We realize that the average velocity of the liquid *within* the medium is the ratio of the *approach* velocity to porosity, or u/ε. And, we realize, the average length of the paths of fluid through the medium is the product of the thickness of the medium, z, and the tortuosity factor, τ.

In the track-etched membrane the tortuosity factor is 1.0, because the pores go straight through.

In a microporous membrane, or a fibrous mat, where the building blocks are arrayed in a random manner, $\tau = 1/\varepsilon$ (ASTM F902). This relationship is derived as follows (Johnston 1992c):

Imagine a plane in a filter medium perpendicular to the flow of fluid. On this plane we draw a grid, where the size of a single square in the grid corresponds to the size of the flow-average pore. Consider that this grid has a thickness corresponding to the length of the sides of the square of the grid. That is, we have a plane of cubes, some empty (the pores), some filled (the filter medium).

The probability, p that a cube is empty corresponds to the porosity. The probability that a cube is occupied (solid) is q, equal to 1 - p. Now consider the flow of fluid through this grid. Consider that N slugs of fluid approach this grid, and the size of a slug equals the size of a cube. We see that Np slugs pass straight through the grid. That is, the distance traveled through the grid equals the thickness.

Of those Nq slugs that hit a solid cube, and have to take a single side step, Npq do find an empty cube, and Nq^2 do not. The total distance traveled by the Npq slugs in passing through the grid is thus twice the thickness.

Of those Nq^2 slugs still on the grid, Npq^2 find an empty cube after two side steps, and thus travel a distance of three times the thickness of the grid in passing through.

Continuing with this logic, the average distance traveled by all slugs, as a multiple of the grid thickness, the tortuosity factor, is

$$\tau = \frac{N(p + 2pq + 3pq^2 + 4pq^3 \ldots)}{N} = \frac{1}{p} = \frac{1}{\varepsilon}$$

In a woven cloth, the value of τ lies somewhere between 1.0 and $1/\varepsilon$. (make your best guess from whether the threads are monofilaments, in which case τ is closer to 1.0; or, are composed of fine fibers, in which case τ is closer to $1/\varepsilon$). For non-woven filter media, consider $\tau = 1/\varepsilon$, from which deduce the flow-average pore diameter, **d**, in a filter medium via (ASTM F 902)

$$\mathbf{d}^2 = \frac{32u\eta z\tau}{\Delta P\varepsilon} = \frac{32B\tau}{\varepsilon} = \frac{32B}{\varepsilon^2} \tag{1.7}$$

The terms are defined in Equation 1.1.

See Section 2.5 in Chapter 2, "Comparing Liquid Flow to Gas Flow."

2
Gas Flow Through Filter Media

2.1 SETUP AND OPERATION OF A GAS-FLOW TEST

INSTRUCTIONS

A common gas-flow test is called the Frazier, in which an upstream, air, gauge pressure of 0.5 lbf/square inch (psi) is applied to a two-inch diameter disk of the filter medium, as the downstream face is exposed to the atmosphere (zero gauge pressure). The air flow from the downstream face is reported as $ft^3/min \cdot ft^2$.

Alternatively, the resistance is reported, say, in mm of water, to an approach velocity of, say, 10.5 ft/min.

In these tests the investigator assumes he or she is in the viscous-flow range. Which is to say that range where the ratio of flow rate to driving pressure is constant for somewhat smaller and larger driving pressures.

Descriptions of setups for panel filters are presented in ASTM F 778.

When the absolute gas pressure (gauge pressure plus atmospheric pressure) on the upstream face of a filter medium is not greater than 10% of the absolute pressure on the downstream face, and, when the *rated* pore diameters are larger than 1.0 μm, then Equation 1.1 (in liquid filtration) can apply. Equations 1.4 and 1.5 can be used to deduce the permeability, and Equation 1.7 can be used to deduce the flow-average pore diameter.

But when the differential pressure is high, and when rated pore diameters are smaller than about 1.0 μm, we employ two other equations to describe gas flow, and we make a plot of the measurements differently than we do for liquid flows.

To perform a gas-flow test with a flat-sheet filter medium, it is convenient to expose the downstream face to the atmosphere, so that we know that the absolute pressure there is one atmosphere. Place the upstream tap for the pressure gauge directly over the upstream face of the medium. Employ a mass-flow meter (such as a rota- or float meter) in the upstream line. Such meters are usually calibrated to read volumetric flow rate at atmospheric pressure. With increasing pressures on the upstream face, record the corresponding increases in gas flow.

Before making the plot we are about to describe, convert the flow rate to the velocity of the gas leaving the downstream face—if you know the surface area. If you don't know the surface area, such as employing a pleated filter medium, then simply record the volumetric flow rate of the emerging gas.

On log/log paper make a plot of the downstream velocity, u_2 (or downstream volumetric flow rate) versus the product of the pressure differential and the average absolute pressures, $\Delta P \cdot \mathbf{P}$, as shown in Figure 2.1.

For filter media rated as having pores *larger* than 1 μm, the plot will look like Line 1. That is, the line starts out with a slope of 1.0, in viscous flow, then falls to 0.5 in inertia flow. (Equation 1.2, in liquid flow, also applies here.) As in liquid flow, we want to employ a filter medium in the viscous-flow range.

For filter media with rated pore diameters *smaller* than about 1.0 μm, the plot will look like Line 2. That is, the line remains straight; however, the slope lies *between* 0.5 and 1.0. What we see in this case is a combination of viscous flow and diffusion or slip flow. Apparently we never reach inertia flow. (Slip flow is not known for liquids.)

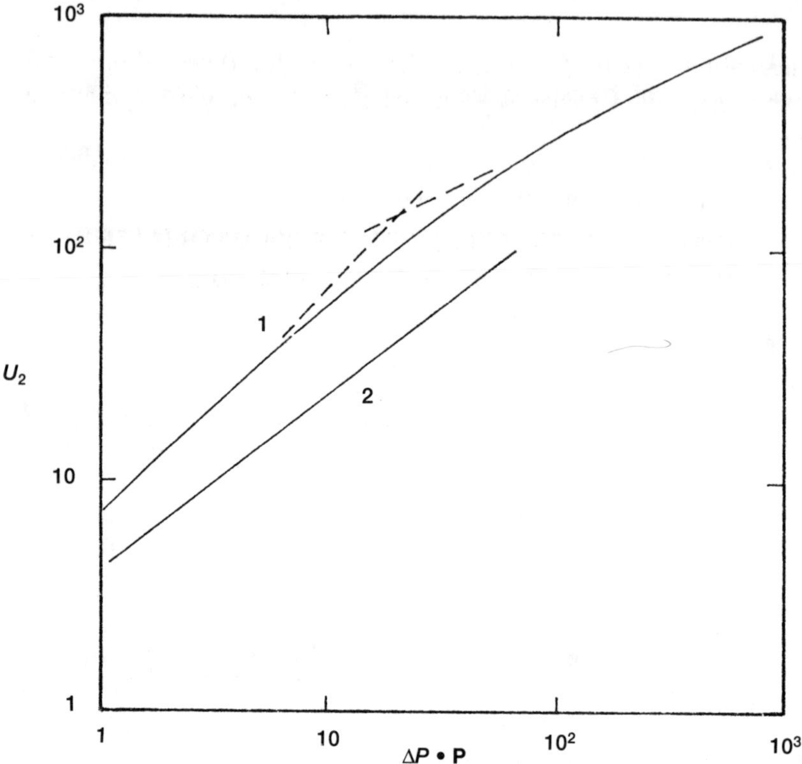

Figure 2.1. Examples of gas-permeability measurements on filter media. The units of measure are arbitrary, but show log/log plots of the velocity of gas, u_2, leaving the downstream face, as a function of the product of the differential pressure, ΔP, and the average of the absolute pressures on both faces, **P**. Line 1 is seen with ordinary media; i.e., the slope starts out as 1.0, in viscous flow, and falls to 0.5, in inertia flow. Line 2 is seen with small-pore media; it remains straight, but with a slope between 1.0 and 0.5, showing a combination of viscous- and slip flow; apparently inertia flow is never reached.

As in the case of liquid-flow tests, be sure the pressure-drop measurements across the medium do not include the pressure drops across the housing. Further, remember that even if the feed-stream pressure is high (as in a compressed-air line) and the filtrate pressure is also high, the items of interest are the *differential* pressure and the corresponding volumetric flow rate from the downstream face.

When the filter medium is soft, be alert that high differential pressures might compress the medium (reduce the porosity) to yield a different medium.

2.2 GAS FLOW THROUGH FILTER MEDIA WITH RATED PORE DIAMETERS LARGER THAN 1.0 μm

BACKGROUND

First, consider Line 1, in Figure 2.1. In those filter media, the description of gas flow is made as follows:

$$\frac{\Delta P\mathbf{P}}{z} = \alpha\eta P_2 u_2 + \frac{\beta(P_2 u_2)^2 M}{R g_c T} \tag{2.1}$$

where: ΔP = absolute pressure on the upstream face, P_1, less the absolute pressure on the downstream face, P_2, N/m^2,

$\mathbf{P}$ = average absolute pressure, $(P_1 + P_2)/2$

z = thickness of the medium, m

α = the viscous-term coefficient, m^{-2}

η = absolute viscosity of the gas, N • s/m^2

β = the inertia-term coefficient, m^{-1}

u_2 = velocity of gas leaving the downstream face, m/s

M = molecular weight of the gas, kg/mole

R = gas constant, 8314 N • m/mole • T

T = absolute temperature, K (°C + 273)

g_c = conversion constant, 1 kg • m/N • s^2.

As in liquid filtration, the rate at which the slope changes is described by Equation 1.2. And as in liquid filtration, our interest lies with the slope-equals-1.0 portion of Curve 1. We are not necessarily interested in the second term. That is, permeability, B, equals $1/\alpha$. And we deduce the flow-average pore diameter, **d**, from

$$\mathbf{d}^2 = \frac{32 u_2 \eta z P_2}{\varepsilon^2 \Delta P\mathbf{P}} \tag{2.2}$$

Equation 2.2 corresponds to the liquid-test Equation 1.7. In Equation 2.2, we assume the tortousity factor, τ is $1/\varepsilon$. Recall the discussion just above Equation 1.7.

2.3 SLIP FLOW THROUGH FILTER MEDIA, WITH RATED PORE DIAMETERS SMALLER THAN ABOUT 1.0 μm.

INSTRUCTIONS

If the plot looks like Line 2 in Figure 2.1, we cannot deduce the viscous-flow-average pore diameter from gas-flow data.

BACKGROUND

In this case we drop the inertia term of Equation 2.1 and replaced it with a slip-flow term. That is, we drop the second term in Equation 2.1, rearrange the equation, then add a second term to obtain

$$u_2 = \frac{\Delta P \mathbf{P}}{z\alpha\eta P_2} + \frac{S\Delta P}{zP_2} \tag{2.3}$$

where S is the gas-diffusion rate, m^2/s, a function of the molecular weight and temperature of the gas, and of the diameters of the pores. The smaller the diameters, the more diffusion, or the more "slip" flow, as we will show by example.

To deduce what portion of the flow is viscous, and what portion is diffusion, we rearrange Equation 2.3 to obtain

$$\frac{u_2 P_2 z}{\Delta P} = \frac{\mathbf{P}}{\alpha\eta} + S \tag{2.4}$$

From this equation, we make a plot on linear/linear coordinates as illustrated in Figure 2.2. Line 1 shows total flow, while Line 2 shows viscous flow. Line 2 is drawn to show zero viscous flow when pressures on

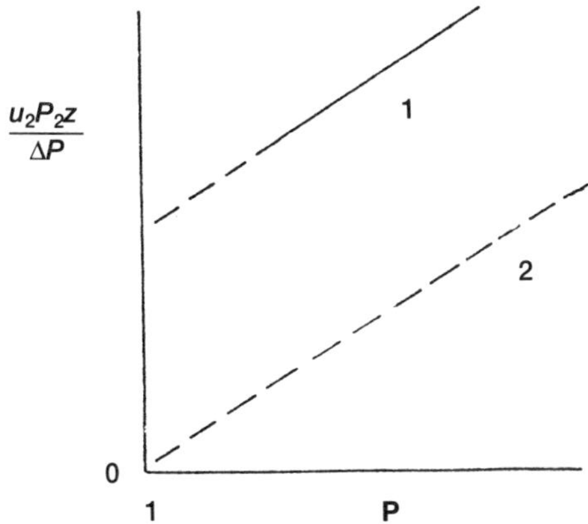

Figure 2.2.. Linear/linear plot of gas-flow data suggested by Equation 2.4. Units of measure on the vertical axis are m^2/s. Line 1 represents total flow (viscous and diffusion). Line 2 represents that portion of the total flow which is viscous flow. In this example, at the point where the average absolute pressure on the separate faces of the filter medium, **P**, is 1.0 atmosphere, the differential pressure, ΔP, is zero, hence no viscous flow occurs. The smaller the pores, the greater the portion of flow that is diffusion flow.

both faces of the medium are equal, that is, when **P** = one atmosphere, or 10^5 N/m^2, and the slope is the same as Line 1.

For a more detailed discussion of S, see Carman (1956) and Badenhop (1983). In general, S is Knudsen flow, and in a capillary tube, the velocity is a function of the diameter of the tube rather than the square of the diameter, as in viscous (Hagen-Poiseuille) flow. We keep these facts in mind when, in Chapter 5, we discuss the extended bubble-point test as a way of measuring the distribution of flow in microporous membranes.

2.4 EXAMPLES OF SLIP FLOW COMPARISONS TO LIQUID FLOW

INSTRUCTIONS

In measuring fluid flow rates through a filter medium with a liquid on the one hand and with gas on the other, we expect to see the same values of permeability. But where the pores are small, and slip flow occurs (and we are not aware that it does), our deduced value of permeability (and flow-average pore size) will be larger with the gas-flow measurements.

An example of what we mean is shown in some fluid-flow measurements with microporous membranes. Figure 2.3 shows data from Millipore Corporation's catalog MC/1, which provides both water and air flow rates, under a common driving pressure, through a series of cellulose–ester membranes. All the membranes have essentially the same porosity and thickness, but the rated pore diameters differ, ranging from 0.05 to 8.0 μm.

Figure 2.3 shows that while we would expect the velocity of both air and water to decline at the same rate with decreasing pore size (all circles will fall on a line with slope of 1.0), we see that, below a certain rated pore diameter, the velocity of gas does not fall as fast as the velocity of water. That is to say, below a rated pore size of near 1.0 μm (the bend in the line), slip flow occurs. And the ratio of slip flow to viscous flow increases with smaller-diameter pores.

2.5 COMPARING LIQUID FLOW TO GAS FLOW

INSTRUCTIONS

Suppose a filter medium shows the symptoms of viscous flow with both a gas and a liquid; that is, the slope of the line in a Figure-1.1 kind of plot is 1.0; and the slope of the line in a Figure-2.1 kind of plot is also 1.0.

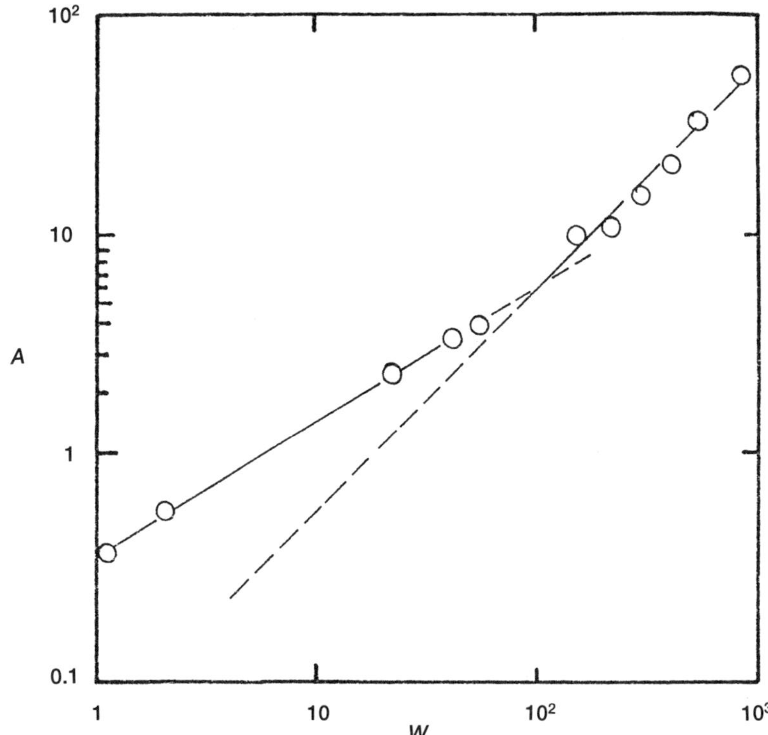

Figure 2.3. Illustrations of slip flow. A = the velocity of air from the downstream faces of membranes, liters/min • cm², versus W, velocity of water, ml/min • cm², when, in each case, the gauge pressure on the upstream face is 13.5 psi. These membranes, of equal porosity and thickness, vary in rated pore diameters from 0.05 μm, on the left, to 8.0 μm, on the right. The straight line changes slope at a point near a rated pore diameter of 1.0 μm. That is, slip flow occurs in filter media with pore diameters smaller than about 1.0 μm.

In this case we can deduce the flow-average pore diameter, **d**, from Equations 1.7 and 2.2; and, we should obtain the same value.

But, suppose we deduce a smaller value for **d** with liquid flow than we do for gas flow, and we are sure that no gas flow is due to slip flow. Further, we are sure that with the liquid-flow test the liquid did not contain enough solids to partially fill, or blind, the pores.

In such cases, it may be that the liquid swells the building blocks of the filter medium, or has some affinity for them. Another liquid may show a different flow-average pore diameter.

Here's a subject for report that we have yet to see: With thick-walled, short, filter cartridges of, say, cellulose on the one hand, and polyolefin on the other, measure the permeability of each cartridge with three separate fluids—air, water, and oil.

3
Building Filter Media From Fibers Or Granules

3.1 PERMEABILITY AS A FUNCTION OF PACKING DENSITY

BACKGROUND

This chapter is directed to the investigator who might estimate (rather than measure) the permeability of a filter medium, and (or) a cake of solids that has collected on the filter medium.

Knowing, or assuming, the total volume of the solids (fibers or particles) in either kind of porous material, V_S, and knowing or assuming the total surface area of those solids, A_S, writers have proposed that the permeability, B, of the mass is proportional to porosity, ε, according to

$$B = \frac{V_S}{A_S} \cdot \frac{\varepsilon^3}{(1-\varepsilon)^2 k}$$

where k is called the Kozeny-Carman constant. The reported value of k is 5, or thereabouts, depending of the general shape of the granules and (or) the assumed shape of the pores. We now show how this expression was derived, showing that it is somewhat misleading. That is, k is not constant, it changes with porosity. Other porosity functions do a better job of describing permeability as a function of packing density.

3.2 THE KOZENY-CARMAN CONSTANT

Where V_o is the volume of void space in a filter medium, and V_s is the volume of the solids, porosity, ε, is

$$\varepsilon = \frac{V_o}{(V_o + V_s)}$$

so that

$$V_o = \frac{V_s \varepsilon}{(1-\varepsilon)}$$

Dividing both sides of the equation by the total surface area of the solids, A_s, we obtain

$$\frac{V_o}{A_s} = \frac{V_s}{A_s} \cdot \frac{\varepsilon}{(1-\varepsilon)} \tag{3.1}$$

Now, separately, consider a tube of internal diameter d, and length L, so that the ratio of internal volume to internal surface area is

$$\frac{V_o}{A_s} = \frac{\pi(d^2/4)L}{\pi d L} = \frac{d}{4} \tag{3.2}$$

Since the surface area of the solids in the filter medium is the same as the surface area of the pores, we combine Equations 3.1 and 3.2 to obtain

$$d = \frac{4V_s}{A_s} \cdot \frac{\varepsilon}{(1-\varepsilon)} \tag{3.3}$$

Here, of course, the diameter of the *average-sized* pore, d, is expressed as a *straight, circular tunnel.*

Now consider that the solids are fibers of diameter d_f and length L_f, so that the volume/surface ratio of an individual fiber is

$$\frac{V_s}{A_s} = \frac{\pi\left(d_f^2/4\right)L_f}{\pi d_f L_f} = \frac{d_f}{4} \tag{3.4}$$

Or, consider that the solids are spheres, of diameter d_{sp}:

$$\frac{V_s}{A_s} = \frac{\pi d_{sp}^3}{6\pi d_{sp}^2} = \frac{d_{sp}}{6} \tag{3.5}$$

From such reasoning, Equation 3.3 can also be written

$$d = \frac{d_f \varepsilon}{(1-\varepsilon)} = \frac{4 d_{sp}\varepsilon}{6(1-\varepsilon)} \tag{3.6}$$

Recall, from Equation 1.7, that permeability, B, is related to the *flow-average* pore diameter, **d**, as

$$B = \frac{u\eta z}{\Delta P} = \frac{\mathbf{d}^2\varepsilon}{32\tau}$$

Carman (1956) relates permeability to the specific d of Equation 3.3 via

$$B = \frac{d^2\varepsilon}{16k}$$

And, substituting the three different expressions of d in Equations 3.3, and 3.6, we write

$$B = \frac{u\eta z}{\Delta P} = \frac{\mathbf{d}^2\varepsilon}{32\tau} = \frac{d^2\varepsilon}{16k} = \frac{V_s^2}{A_s^2} \cdot \frac{\varepsilon^3}{(1-\varepsilon)^2 k} \tag{3.7a}$$

$$= \frac{d_f^2}{16} \cdot \frac{\varepsilon^3}{(1-\varepsilon)^2 k} \tag{3.7b}$$

$$= \frac{d_{sp}^2}{36} \cdot \frac{\varepsilon^3}{(1-\varepsilon)^2 k} \tag{3.7c}$$

Notice, k, incorporates the tortuosity factor τ, which changes with porosity (see Section 1.9); but, that fact is not embodied into these Equations

3.3 THE KOZENY-CARMAN CONSTANT FOR FIBERS

For fibrous media, the correlation between ε and k, in Equation 3.7b, is seen in Table 3.1 (Rushton and Griffiths 1977).

Table 3.1
Values of *k* in Equation 3.7b
for a Random Array of Fibers

ε	0.2	0.3	0.4	0.5	0.6	0.7	0.8	0.9
k	2.7	3.8	4.9	5.8	6.3	6.6	7.2	9.8

When Monson (in ASTM's STP 975, Vol. I, pp. 27-45) looks at the permeability of a fibrous mat he does not address the Kozeny-Carman constant. Instead, he calculates the viscous drag of a fiber as it is held perpendicular to the path of the fluid. He then goes on to consider many fibers, randomly arrayed in a fibrous medium and packed to a solidity, c. Recall, $c = 1 - \varepsilon$.

In describing the (dimensionless) viscous drag of the medium, F, as a function of c, (verified by actual measurements), he relates these values to permeability (in the present notations) as

$$B = \frac{u\eta z}{\Delta P} = \frac{\pi d_f^2}{4cF} \tag{3.8}$$

His charts provide the data for Table 3.2.

Table 3.2
Values of *cF* in Equation 3.8, as a function of *c*

c	0.9	0.8	0.7	0.6	0.5	0.4	0.3	0.2	0.1	0.05
cF	25,000	4,400	1,050	330	125	48	16	4.8	1.3	0.38

Using the data in Tables 3.1 and 3.2, we draw the curve, in Figure 3.1, of B/d_f^2 versus ε.

Two mathematical expressions describe separate portions of the curve of Figure 3.1:

- In the range of $c = 0.006$ to 0.3, i.e., $\varepsilon = 0.994$ to 0.7 (Davies 1973):

$$\frac{B}{d_f^2} = \frac{1}{[64c^{1.5}(1+56c^3)]} \tag{3.8}$$

- In the range of $\varepsilon = 0.1$ to 0.85 (Johnston 1989):

$$\frac{B}{d_f^2} = 2.66 \cdot 10^{-3}\left[\frac{\varepsilon(1+\varepsilon)}{(1-\varepsilon)}\right]^2 \tag{3.9}$$

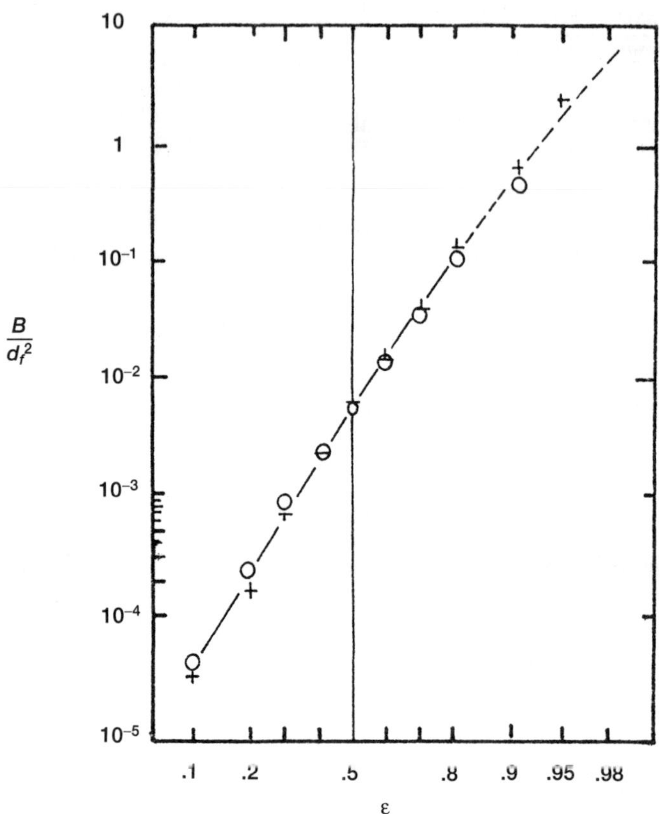

Figure 3.1. Permeability, B, of a mat of fibers of diameter d_f, expressed as the ratio B/d_f^2, as a function of porosity, ε. Points **o** are from Table 3.1, and points + are from Table 3.2. The broken line is described by Equation 3.8, the solid line by Equation 3.9. Recall, $c = 1 - \varepsilon$.

3.4 THE KOZENY-CARMAN CONSTANT FOR GRANULES

We have not seen the clear-cut values of k reported for granules as we have seen for fibers. When Carman (1937) looked at the permeability of granular beds, with porosities, ε, ranging from 0.37 to 0.66, he saw k values ranging from 4.8 to 6.13. He made no attempt to correlate one value with the other. And while he reports that the tortuosity factor in a packing of spheres is $\sqrt{2}$, he does not use that thinking in concluding that, on the average, *k* values, in granular beds, are 5 (not the more precise 5.0 that some writers have reported).

Ergun (1952) looked at similar data and, while assuming that the volume/surface ratios of the granules are those of an "equivalent sphere" (Equation 3.7c), concludes that, on the average, $k = 4.16$.

Macdonald et al. (1979), looking at the viscous-flow term of Ergun's equation (the present Equation 3.7c), offer that ε^3 should be replaced by $e^{3.6}$. Which is to say that *k* is indeed a function of ε. Thus, we gather, $k = 4.16/\varepsilon^{0.6}$.

On the other hand, Meyer and Smith (1985) offer that the ε^3 term in Ergun's equation should be replaced by $\varepsilon^{4.1}$, leaving us to gather that $k = 4.16/\varepsilon^{1.1}$.

Dullien (1979), in reviewing the work of a dozen authors, concludes that with beds of granules, we cannot characterize the permeability of that bed by the two parameters volume/surface ratio, and the porosity function of Equations 3.7a–c, that is, the function $\varepsilon^3/(1-\varepsilon)^2$. He lists other porosity functions investigators have suggested; but, none provide a general formula for all granular shapes.

Indeed, how do we measure, or assume, the volume/surface ratio of irregularly shaped granules?

4
Concepts Of Pore-Size Distributions

BACKGROUND

As a prelude to Chapter 5, where we discuss the methods and results of measuring pore-size distributions, the present chapter discusses what kind of results we expect to see.

4.1 A MECHANICAL MODEL OF PORE-SIZE DISTRIBUTIONS

Imagine this experiment: Randomly drop sticks pins or toothpicks onto a flat surface. Take a photograph of the array (drops the sticks onto the glass of a copy machine). With a planimeter, measure the surface area of each of the open-space polygons on that photograph. Deduce the "radius" or "diameter" of each polygon to be the square root of the area (remember, we are dealing with relative values).

On following these steps we find a number distribution of pore diameters, that :

- closely follows a log-normal distribution, and
- has a geometric standard deviation of about 2.2.

Further, no matter the number of sticks we employ, it appears that the distribution does not change (Piekaar and Clarenburg 1967); but, as one might expect, the more sticks, the smaller the number-average pore diameter. To say this in another way, no matter the density of the sticks:

The ratio of the diameter of the "largest" pores to the diameter of the "smallest" pores remains the same.

We will keep this conclusion in mind when we do the measurements outlined in Chapter 5. Further, as we easily reason, for a given mass (weight) of sticks per area, the use of small-diameter sticks yields smaller pores than does the use of large-diameter sticks, because, for equal masses of sticks, we have more numbers of small sticks than large ones.

Thus, we further reason: Given fibers of a certain diameter, the more closely we pack them into a sheet (that is, by decreasing the porosity or increasing the solidity), the smaller the pores.

We have yet to see an investigator do the above experiment with irregularly shaped granules. We hope a reader might do it (say, with crushed stone) and report the results, to answer the question arising in Section 4.4:

Is the pore-size distribution in a bed of granules narrower than the pore-size distribution in a bed of fibers of equal porosity?

4.2 FROM *FLATLAND* TO THREE DIMENSIONS

To extend the above *Flatland* (Abbott 1963) view of a pore-size distribution to a distribution of three dimensions, consider that a fibrous filter medium is composed of many theoretical, thin layers. The pore-size distribution in one layer is the same as in the next layer; however, a large pore in one layer does not necessarily lie in the same spot on the next layer. Thus, once a fluid penetrates the first layer, an individual stream, on meeting the next layer, may either divide into smaller streams, or combine with adjacent streams to make a larger stream, and the cycle is repeated as the fluid goes on to the next layers.

In the case of the sponge-like, microporous membrane, we can view the structure as a packing of the skeletons of polyhedra, resulting from closely packed bubbles that have dried out. Where two skeletons touch with a common face, the opening in that face is a pore. The diameters of the polyhedra are larger than the diameters of the pores, and of course there exists a distribution of both kinds of diameters.

In membranes produced by stretching a solid sheet of plastic to yield a fiber-like network, or in beds of particles, we envision that the fluid-flow path is no different in principle than what we describe above.

4.3 A *FLATLAND* MATHEMATICAL MODEL OF PORE-SIZE DISTRIBUTIONS THE MOST PROBABLE DISTRIBUTION

Consider placing a grid over a plane in a filter medium, which medium has been built of a random array of building blocks. Consider that the size of an individual square in the grid corresponds to the size of the smallest pores in the medium. The probability that an individual square on the grid covers a pore, or part of a pore, is ε, the porosity of the filter medium.

The probability that an adjoining grid square is part of the same pore is also ε. Thus, the probability that a pore has a diameter, or radius (some linear dimension) of x grid steps, is ε^x. Hence, an expression for the (relative) numbers, N, of pores of diameter x is

$$N_x \sim \varepsilon^x x \tag{4.1}$$

Now consider that the cross-sectional area of the pore corresponds to x^2, and that the pore has unit depth. It then follows that the volume distribution of pore sizes is described by

$$V_x \sim \varepsilon^x x^2 \tag{4.2}$$

Further, since the volumetric flow rate of fluid, through a pore, under a given driving pressure, must be proportional to the square of the cross sectional area of the pore (Hagen-Poiseuille flow, Equation 1.6), the distribution of the volumetric flow rate, Q, through a pore of diameter x must be described by

$$Q_x \sim \varepsilon^x x^4 \tag{4.3}$$

Where the pores are small enough so that slip flow occurs (with gas flow), the flow distribution apparently lies with an adaptation of Equation 4.3 wherein the exponent over x is less than 4.

To evaluate each of these three equations we employ a computer to calculate the values for N, V, and Q, for increasing values of x, carrying ten-digit numbers until we reach a converging sum. The *average* value of x in each of these distributions is reached via the kind of calculation seen in deducing the average value of x in Equation 4.3

$$\text{Average } x \text{ value in Equation 4.3} = \frac{\Sigma(\varepsilon^x x^5)}{\Sigma(\varepsilon^x x^4)} \tag{4.4}$$

We now put these equations to use. Consider a bed of granules with a porosity, ε, of 0.35. Here the pore-volume distribution (deduced via fluid-intrusion measurements) should be shown by Curve A of Figure 4.1, and the fluid-flow distribution should be shown by Curve B. Circles on these curves shown the average pore diameters in each distribution.

In order to plot these curves as *cumulative* functions, we, with a computer, deduce the value of $f(x)$ for x values at intervals of 0.1, extending the calculation for increasing values of x until we reach a converging sum. After normalizing that sum, $\Sigma = 0.999\ 999\ 999\ 9$, we make the plots shown in Figure 4.2 on log-probability paper.

Figure 4.2 also shows the same kinds of distributions where porosity is 0.75. If these curves were straight lines we would see true log-normal distributions, with geometric standard deviations in the range of 1.5–1.8. (To deduce the value of the geometric standard deviation, notice the values of x where Σ, on the left-vertical scale is 0.50. Notice the value of x where $\Sigma = 0.84$. Divide the latter x value by the former.)

Thus the curves in Figure 4.2 amount to approximatly-log-normal distributions—the kinds we expect from the mechanical model of Section 4.1. In such distributions, the three different kinds of distributions, i.e., the *number*, the *volume*, and the *fluid flow*, all have similar, geometric standard deviations (while, of course, the averages are different).

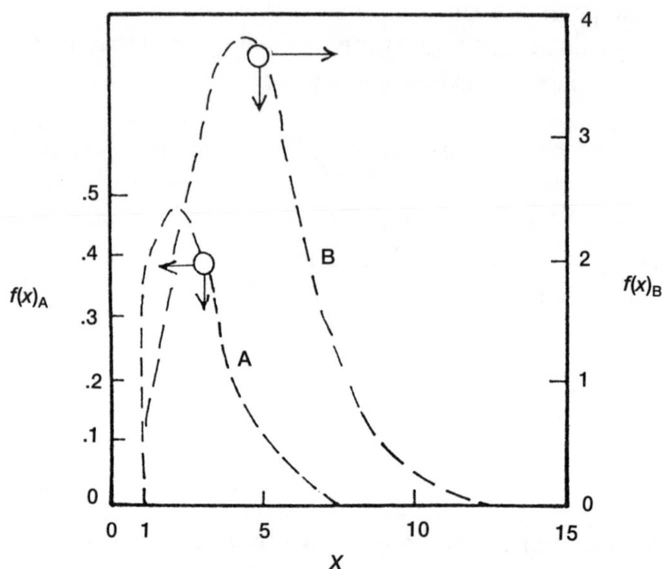

Figure 4.1. Curve A, showing $f(x) = 0.35^x x^2$, is the most probable pore-size disribution according to the volumes of pores of diameter x, from Equation 4.2, when porosity, ε, is 0.35. Curve B is the distribution of fluid flow, $f(x) = 0.35^x x^4$, from Equation 4.3. Circles on each curve show the average pore diameter, where the smallest pore diameter is 1.0 unit.

The geometric standard deviations in Figure 4.2 are somewhat near to what we expect from the mechanical model of Section 4.1.

4.4 ANOTHER MATHEMATICAL MODEL OF A PORE-SIZE DISTRIBUTION

This mathematical model yields a narrower pore-size distribution. It is suggested by the results of liquid-drainage tests performed on a bed of spheres (see Chapter 5). The volume distribution of pore diameters, comparable to Equation 4.2, is

(Text continued on page 34)

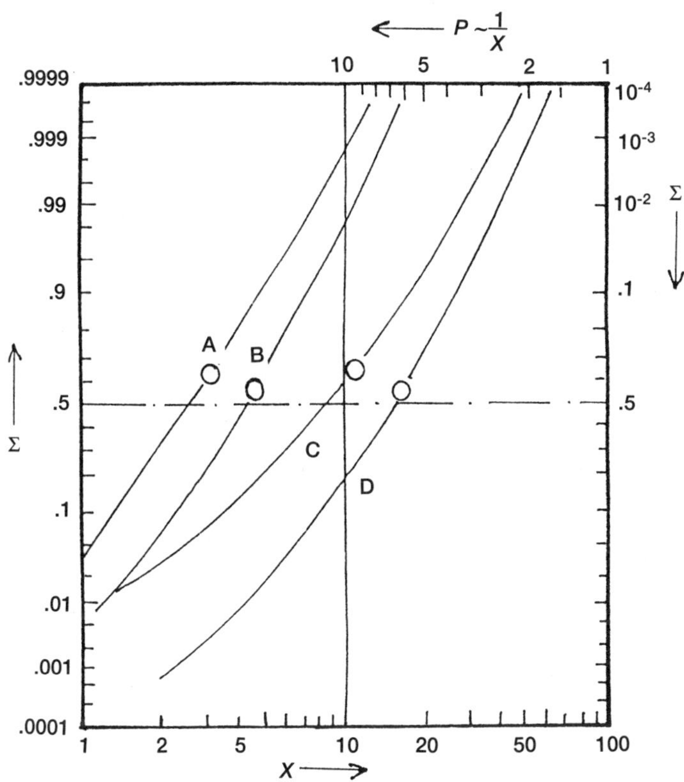

Figure 4.2. Cumulative, most-probable, pore-size distributions, deduced from Equations 4.2 and 4.3. Curve A is the volume distribution of pore diameters, *x*, and Curve B is the fluid-flow distribution, from Figure 4.1, where porosity, ε, is 0.35. Curve C is the volume- and Curve D is the flow distribution where porosity is 0.75 (the porosity of microporous membranes and many fibrous media). Circles on the curves indicate the average pore diameters. The upper right hand corner, addressed in Chapter 5, shows the relative amounts of increased gas pressure, $P \sim 1/x$ (from Figure 5.1), necessary to blow liquid from saturated media, in following Curve D or B.

(*Text continued from page 32*)

$$V_x \sim x^4(1-x)^2 \tag{4.5}$$

In this distribution the smallest pore has a diameter of zero, and the largest pore 1.0. The cumulative-sum distribution is shown in Figure 4.3 as Curve 2. The volume-average pore diameter is 0.625, which can be deduced from Equation 4.7.

In calculating the cumulative sums of Equation 4.5, so that we can draw the curve in Figure 4.3, we determined values for V_x at x-value intervals of 0.02, added them together, and then normalized the sum.

Equation 4.5 is a form of the Beta function (a mathematician's term, not to be confused with the Beta ratio in Chapter 7):

$$f(x) = Ax^a(1-x)^b \tag{4.6}$$

where A provides that the area under the curve is unity. The average value for x is deduced as follows:

$$avg.\ x = \frac{a+1}{a+b+2} \tag{4.7}$$

If Equation 4.5 describes the volume distribution of pore diameters, then (via the reasoning in Section 4.3 with the most probable distribution) the number distribution in a bed of spheres must be described by

$$N_x \sim x^2(1-x)^2 \tag{4.8}$$

which is shown in the cumulative form in Figure 4.3 as Curve 1.

By the same reasoning, the fluid-flow distribution must be described by

$$Q_x \sim x^6(1-x)^2 \tag{4.9}$$

that we show in Figure 4.3 as Curve 3.

Haring and Greenkorn (1970) suggest the use of this Beta function by way of their results in an air-pressure-induced, liquid-drainage test on a bed of spheres. Their drainage data, as well as other drainage data (Bear

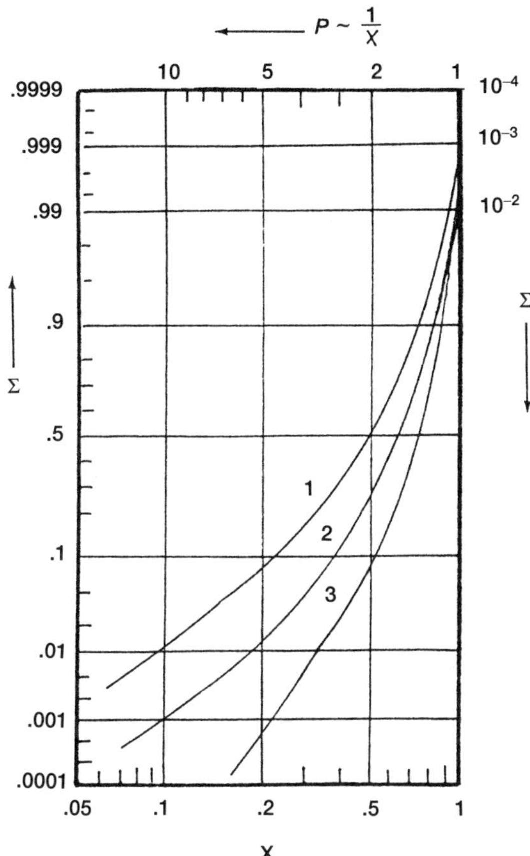

Figure 4.3. Plots of cumulative pore-size distributions according to the Beta function. Pore diameters are *x* units. Curve 1 is from Equation 4.8; Curve 2 from Equation 4.5; and Curve 3 from Equation 4.9. Curve 2, from the upper right corner, corresponds to the increased pressure of gas, ($P \sim 1/x$) on a liquid-saturated bed of spheres, to push liquid from the bed, but not enough to blow pores open and allow gas flow.

1972), correspond to Curve 2 of Figure 4.3, which curve, in its differential form, is described by the exponents in Equation 4.5.

It would seem that such liquid-drainage data describe a *volume* distribution of pore sizes. Yet, Haring and Greenkorn maintain that their model, with the exponents in Equation 4.5, represents a *fluid-flow* distribution.

5

Fluid-Intrusion Measurements The Pore-Size Distribution The Bubble Point and the "Largest" Pores

5.1 INTRODUCTION

BACKGROUND

As a general rule, the filter user may not be interested in or concerned with the subject matter in this chapter, unless the filter medium is used in a critical step (for example, as a sterilizing filter in a pharmaceutical operation). Outside of such an operation, the filter user already has a feel for the flow-average pore diameter if he or she has studied the fluid-flow data as presented in Chapters 1 and 2, or has been told the flow-average pore diameter. Thus, having made the best estimate of which pore size would be best for the fluid to be filtered, he or she simply proceeds with a filtration test, or a full-scale run.

However, when a *fine* filter medium is required in a critical filtration step, the filter user will want (or, indeed be required) to test the installed filter medium for integrity and bubble point before starting the run. The integrity test is tied to the bubble-point test, which we address first in this chapter.

An extension of the bubble-point test, as a separate test in the laboratory, leads to the determination of the "mean-flow" pore size. What that test is really attempting to measure (as we describe below) is the *median* (middle, not average) pore size. The ASTM committee with jurisdiction over test method F 316, that describes how to make such a measurements, perpetuates the *mean*-flow misnomer. And investigators, making reference to that test method simply follow that nomenclature without comment.

Other laboratory tests that attempt to determine the pore-size distribution are also described in this chapter, namely, the mercury-intrusion test, and the liquid-drainage test.

5.2 THE INTEGRITY AND THE BUBBLE-POINT TESTS

At first glance, the principles behind these tests are quite simple: On soaking a horizontally held filter medium with a liquid, making sure to flood all pores, and leaving a pool of liquid on top, air is applied underneath, with slowly increasing pressure. As the air reaches a certain pressure, the largest pores blow open; and, we see bubbles of air emitting from the top side.

Thus, *bubble point* means the *pressure* that causes the first bubbles.

When we know beforehand what the bubble point might be, the failure to see any bubbles at lower pressures tells us that we have installed the medium in the housing with *integrity.* That is, we have no apparent leaks around the seal, and we have no tears or holes in the medium.

Because the sizes of the pores are inversely proportional to the bubble pressure (explained in the next section), we look for a pressure that is high enough. If the bubble point is too low, we assume one of the following:

- We have the wrong filter medium in place (it has large pores).
- Even if it is the correct medium, it must have a tear or other holes.
- The medium is not well sealed in the housing; the filter medium has no *integrity.*

Because the determination of the bubble point is employed to deduce the size of the "largest" pores, we now explain the reasoning behind that deduction, along with common errors investigators make.

5.3 THE PRINCIPLE OF THE BUBBLE-POINT TEST

As stated above, this test is based on the principle that when a filter medium, e.g., a 47-mm-diameter disk, is placed in a housing and covered with a liquid (saturating the medium), and a gas pressure is slowly increased from underneath, the gas will push liquid from the largest pores first, with the result that bubbles of gas will appear on top of the medium, under the cover of the liquid layer (ASTM E 128, F 316). The higher the pressure to cause this first bubble, the smaller these "largest" pores. This inverse relationship is explained in Figure 5.1.

Even though we may know the surface tension of the liquid employed to saturate the medium, there's always the question: Has material within the medium changed the surface tension? To address this question, we flush enough liquid through the medium to clean it out, while perhaps measuring the surface tension of the effluent liquid.

In deciding what value to use for the wetting angle, α, we assume, in the force vector, $\gamma \cos \alpha$, of Figure 5.1, that $\cos \alpha = 1.0$. With the exception of the track-etched membrane, a pore is not (of course) the neat, circular tunnel depicted in Figure 5.1, with a smooth and well-defined wall; and we have not yet defined a probability-average force vector along the rough walls of a pore.

Some investigators include a *correction factor* in the equation of Figure 5.1, so that the calculated value of r (pore radius) will correspond to some "correct" value. That action brings to mind the classical tale:

> In a small town before the time of radio, citizens set their watches by either the big clock at the bank or the factory whistle. When a newspaper reporter asked the banker how he knew the correct time to set his clock, he replied, "from the factory whistle." When the man at the factory was asked how he knew the correct time to sound the whistle he replied, "from the big clock at the bank."

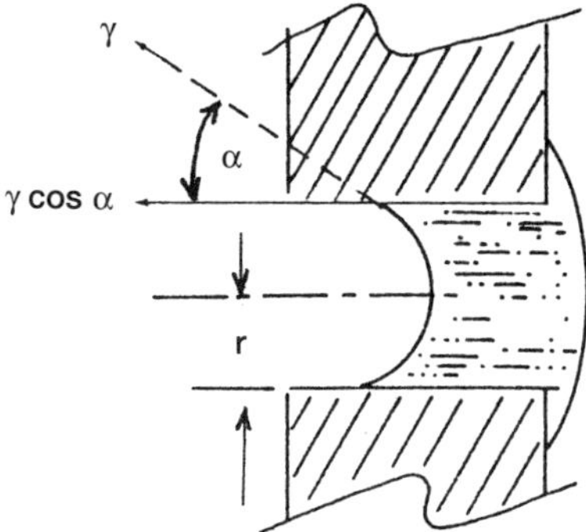

Figure 5.1. Illustration of how gas, from the left, and pressing toward the right, forces a liquid from a pore and how the required pressure of the gas, P, is inversely proportional to the radius of the pore, r. In this example, the liquid wets the pore wall (here a cylinder) at an angle α, while the cosine vector of the surface tension, γ, holds the liquid in place. The expression for these balances of forces is

$$P = \frac{\text{perimeter force}}{\text{cross-sectional area}} = \frac{2\pi r \gamma \cos\alpha}{\pi r^2} = \frac{2\gamma \cos\alpha}{r}$$

where: P = gas pressure, N/m^2
γ = surface tension of the liquid, N/m
α = wetting angle
r = pore radius, m

5.3.1 The History Of The Bubble-Point Test

In determining the bubble point for membranes, the test described at the beginning of Section 5.3 was first performed on 47-mm-diameter samples. When the test was subsequently performed on 293-mm-diameter samples, the investigator did not look at a pool of liquid over the membrane. Instead he employed a setup whereby the gas flowing from the membrane is funneled into an eye dropper held under water. He then looked for the first continuous flow of bubbles emerging from the eye dropper.

But a funny thing happened. He saw bubbles flowing from the eye dropper at gas pressures lower than he expected from the test results with a 47-mm-diameter sample.

Did this mean he failed to properly seal the 293-mm disk in its housing? Or did it mean his large-diameter sample contained a tear, or pinholes? The explanation came when Reti (1977) realized that with a 293-mm-diameter membrane, he saw a flow of air that was simple diffusion of dissolved air through the liquid in the membrane before any pores are blown opened. With the 47-mm sample, the volumetric flow rate of diffused air, at the low pressure, is too small to see.

This diffusion is not the slip-flow type addressed in Section 2.3. That is, on the high-pressure face of the membrane, gas is forced into solution in the liquid. The dissolved gas then diffuses to the low-pressure face to come out of solution and give the *appearance* of hydrodynamic flow though pores.

That this flow of gas is indeed diffusion is deduced from the following:

- The flow rate is consistent with calculations regarding the pressure-dependent solubility of gas, the thickness of the liquid (thickness of the membrane), the cross-sectional area of the water (porosity of the membrane), and the diffusion rate of dissolved gas from one liquid face to the other.
- In a log/log plot of gas flow rate versus driving pressure, the slope is 1.0, as shown by Line B in Figure 5.2. If the slope were higher, as Curves X, Y, or A, we would see the symptom of hydrodynamic flow.

5.3.2 Calculating the Flow Rate of Diffused Air

To deduce this air-diffusion rate, we employ the procedure taught by Treybal (1980). In the following example we consider:

- room temperature,
- air pressure on Side 1 is 2 atm absolute,
- pressure on Side 2 is 1 atm absolute,
- thickness, of a microporous membrane, 136 μm.

The basic equation is

$$N_A = \frac{D_{AW}\rho(x_{A1} - x_{A2})}{zyM}$$

where: N_A = flow of dissolved air in water, kmol/m$^2\cdot$ s
D_{AW} = diffusivity of air in water, $1.6 \cdot 10^{-9}$ m^2/s
ρ = average density of the mixture, 1000 kg/m^3
M = average molecular weight of the mixture, 18 kmol/kg
x_{A1} = mole fraction air in mixture on Side 1 = 0.0000278
(Henry's constant is $7.2 \cdot 10^{4.}$ Thus,
$x_{A1} = 2/7.2 \cdot 10^4 = 0.0000278$.)
x_{A2} = mole fraction air in mixture on Side 2 = 0.0000139
z = thickness of the water (the membrane), $135 \cdot 10^{-6}$ m
$y = (x_{w2} - x_{w1})/\ln(x_{w2}/x_{w1})$, w referring to water concentration on either face of the mixture.

Thus,
$y = (.9999722 - .9999861)/-1.39\cdot10^{-5} = 1.00$, and
$N_A = 1.6\cdot10^{-9}\cdot 1000 \cdot 1.39\cdot10^{-5}/135 \cdot 10^{-6}\cdot 18 = 9.15 \cdot 10^{-9}$ kmol/m$^2\cdot$s.

Finally, with the porosity of the membrane at 0.75, the velocity of air emitting from the low-pressure side is
$9.15\cdot10^{-9}$ (kmol/m$^2\cdot$ s) $\cdot$ 24.6 (m^3/kmol) $\cdot$ 0.75 = $1.69\cdot10^{-7}$ m/s.

5.3.3 Accurate Measurement of the Bubble Point

To obtain a precise measurement of the bubble point we must also actually measure the flow rate of air. For example, in the case of the 293-mm-diameter membrane, simply viewing the first bubbles is not enough. In one experiment, seven people were separately asked to indicate when each first saw a steady flow of air through an eye dropper held under water. Responses varied from 5 to 50 ml/min, revealing that one person's perception of first-steady flow is ten times that of another's.

It is important to view a log/log plot of gas flow rate versus driving pressure such as that shown in the "generic" example of Figure 5.2.

(*Text continued on page 43*)

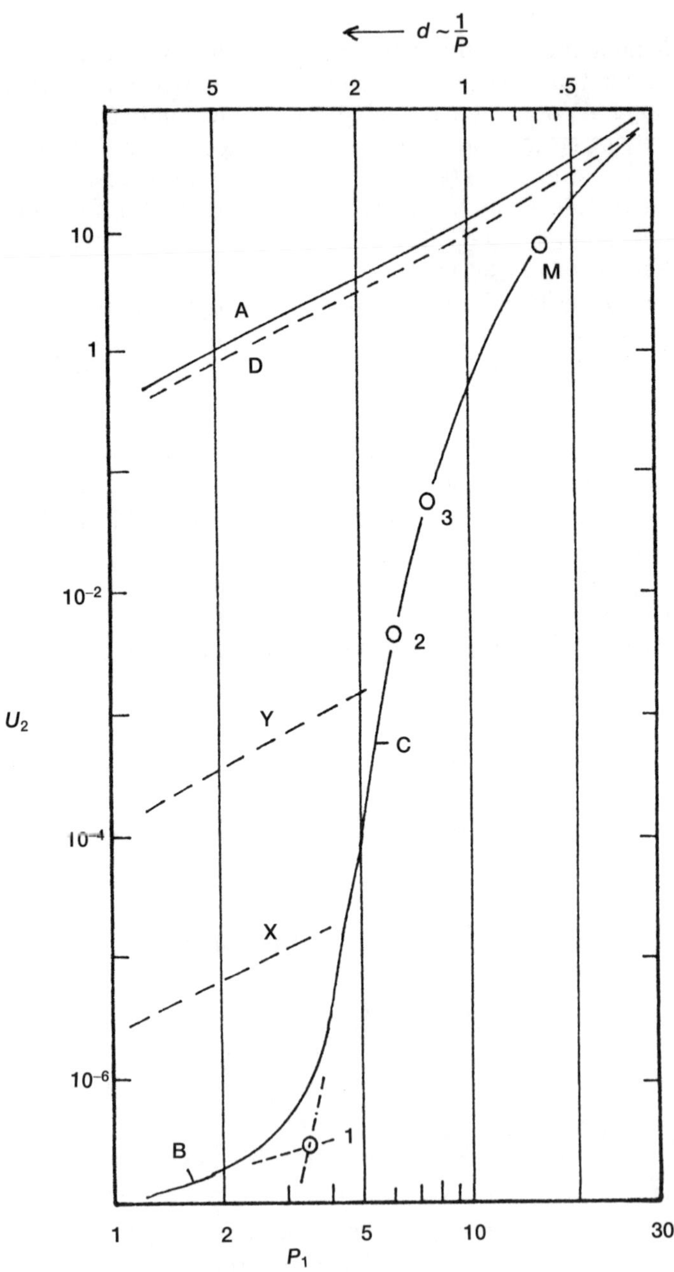

Figure 5.2 (*Caption on facing page*)

Figure 5.2.(*Shown on facing page*) Measuring the bubble point, Point 1, an indication of the "largest" pore, and measuring how fluid flow is distributed among the pores, by comparing Curve C to Curve D.

u_2 = velocity of air leaving the downstream face of the filter medium at zero gauge pressure; P_1 = gauge pressure on the upstream face; $d \sim 1/P_1$ = pore diameter. Units of measure for u_2 , P_1, and d are arbitrary, but the relative shapes and positions of these curves are essentially universal in this plot on log/log paper.

Curve A shows air flow from a dry medium. Curve B is the flow of air diffusing through the liquid-saturated medium, before any pores are blown open, after which, with increasing air pressure, we see the fast-increasing Curve C, which never quite reaches Curve A (in correctly employing a liquid that does not evaporate). Curve D (flow through a damp medium) is traced (with decreased pressure) after establishing Curve C.

Point M represents the flow-*median* pore diameter (M is half the u_2 value of Curve A, or D (A is close to D). Curves X and Y (two separate cases) represent air flows when extra-large openings have been blown free at pressures below P_1 = 1.0 unit.

The true bubble point is Point 1. Less-sensitive air-flow measurements indicate the bubble point to be, say, Point 2, or Point 3.

(*Text continued from page 41*)

"Generic" means the units of measurement are arbitrary but the shapes and positions of the curves are real (a convenient feature of log/log plots).

Curve A shows the velocity of air leaving the downstream face of a *dry* membrane as a function of driving pressure. In this kind of plot, Curve A turns up. It would be a straight line if the horizontal axis were expressed as $\Delta P \cdot \mathbf{P}$, as in Figure 2.1.

Line B shows the flow of air through a liquid-saturated membrane (via diffusion, as discussed in Section 5.3) before any pores blow open.

Curve C shows the fast-increasing flow of air after the first pores are blown open. The bubble point is Point 1 (ASTM STP 975, Vol II, pp. 59–68; Johnston 1992c) where, in this example, P_1 = 3.5 units.

Curves X and Y represent two separate cases in which many pin holes (or a single large hole) blew open at a pressure less than P = 1 unit.

5.3.4 Extension of the Bubble-Point Test

With increasing pressure, Curve C never really reaches Curve A. That is, when the procedure correctly employs either a liquid of low vapor pressure or a gas presaturated with the liquid (to avoid evaporative drying of the pores), the action of blowing open all the pores still leaves some liquid along the pore walls. In that case, the flow of air is not as great as it is through a dry membrane at the same driving pressure.

The data necessary to draw Curve D is obtained after taking the data for drawing Curve C. That is, Curve D represents flow through a "damp" membrane with all the pores blown open.

As mentioned above, the true bubble point is the extension shown as Point 1. That is to say, where we use this kind of bubble-point measurement to deduce the size of the "largest" pore, we see that, in this example (taken from data around 0.45-μm-rated membrane wet with water and using air to blow open the pores) (Johnston 1992c), the largest-diameter pores we can reach (in the absence of Curves X and Y) account for about 10^{-7} of the total fluid flow (*not* that fraction of the numbers or the volumes of the pores).

We can't account for any more of the distribution unless we employ a liquid that dissolves less of the gas used.

5.3.5 Defining the Diameter of the "Largest" Pores

Accounting for 0.999 999 9 of the total flow (seven nines) does embrace enough of the pore-size distribution for the pharmaceutical industry, which is interested in separating microbes from fluids. But in other applications, and with other filter media, it may not be necessary to account for so much of the distribution.

It may be sufficient to account for, say, only 0.999 of the distribution in defining the "largest" pore. In this case our apparatus need only measure the first flow of air under a given driving pressure that is, say, 0.001 of that through a dry or damp membrane, at Point 2, where P_1 = 5.5 units and (top scale) where the pore diameter is 1.82 units.

Alternatively, we may be interested in only accounting for 0.99 of the distribution, that is, where the bubble point corresponds to an air flow of

0.01 of that through a dry membrane, at Point 3, where $P_1 = 6.8$ units, and the pore diameter is 1.47 units.

Obviously these two differently perceived diameters of the "largest" pores differ by 24%; and our choice of deciding how sensitive an air flow meter to use rests with our choice of what we really want to measure.

The investigator-writer worth his or her salt will tells us how much of the pore-size distribution has been accounted for when he or she reports the bubble point.

Point M represents the pressure, and corresponding pore size, where half the pores (defined by fluid flow) have been blown open. Thus, in this example, the *median*-flow pore diameter is 0.65 units. Many investigators, following the teachings of ASTM F 318, call this the "mean flow pore."

5.4 THE INTEGRITY TEST

The pharmaceutical industry has the requirement that a filter medium, once in place, must be tested for integrity and bubble point before one proceeds with the filtration run.

The integrity test consists of applying an air pressure to a previously liquid-saturated membrane at a level lower than the bubble point (e.g., in Figure 5.1, at $P_1 = 2$ units) in order to look for diffusion flow, and to measure that flow rate. If that flow rate is no greater than expected (see Section 5.3.2), then proceed with the test by increasing the air pressure to look for the bubble point.

Various instruments are available to perform these tests automatically. Some even provide a printout for the required record of production (Johnston 1992c). But suppose in a pharmaceutical application, where $P_1 = 1.0$ unit, we see an air flow indicated by Curve X instead of Curve B. Here, as a worse-case scenario, we might assume that during filtration, when microbes approach the membrane, perhaps 10^{-5} of their number will stream through unstopped, so that filtration efficiency is 0.999 99, instead of the required 0.999 999 9.

However, this scenario will not be realized. In many microbe-filtration experiments with membranes containing such flaws, the efficiency of filtration did not suffer (ASTM STP 965, Vol. II, pp. 27–36). We assume filtration efficiency *might* suffer, however, if we see data such as Curve Y.

In the case of a coarse filter medium, the largest pores may be too large to hold the liquid. Alternatively, the liquid may have a low surface tension. In either case the liquid simply falls out under the force of gravity, and we will see the types of curves indicated by Curves X and Y. In these cases our measurement of the bubble point cannot be lower than that indicated by the example of Point 2.

5.5 DISTRIBUTION OF FLUID FLOW

By comparing Curve C to Curve D in Figure 5.2, we deduce the magnitude with which different-sized pores carry portions of the total flow through the filter medium. This writer understands that the Coulter Porometer™ operates in such a manner. Some investigators compare Curve C to Curve A. And some investigators fail to address the fact that some pores may have been emptied by evaporation.

At the pressure reading on Curve C where the gas flow rate is half what it is through a dry membrane, Curve A (or damp membrane, Curve D, depending on the operator) corresponding to Point M, investigators deduce a pore diameter they call "the *mean* flow pore," presumably because they hold on to that misnomer used in ASTM F 316 (as we mentioned in Section 5.1). But, of course, what the procedure really attempts to measure is the *median*-flow pore diameter. The mean is the average; the median is the middle.

Yet in a sub-μm-rated membrane, portions of the air flow described by Curves A, C, and D, are slip flow (discussed in Chapter 2), which raises the question: By comparing Curve C to A, or to D, are we really seeing the comparison we would see if all flows were viscous flow—the kind of flow experienced in liquid filtration? Recall, from Chapter 2: Under viscous conditions, the *velocity* of a fluid through a pore, under a given driving pressure, is proportional to the square of the diameter, whereas in slip (Knudsen) flow velocity is proportional to the diameter.

Further, when the filter medium contains large pores, so that no slip flow occurs along Curves C, A, and D, but high pressures are still required to blow liquid from the pores, we face another question: Are those air flows in the inertia-flow region, instead of the viscous-flow region? In inertia flow we don't know how fluid flow is distributed among the pores. We don't understand the meaning of β in Equation 2.1.

By employing a liquid with a lower surface tension we can reduce the pressures, and perhaps keep the air flows in the viscous range. In any event, the diligent investigator will examine Curve A in the light of Figure 2.1 to determine what kind of air flows occur.

Thus, the determination of the "mean-flow pore," and, indeed, the bubble point, as directed by ASTM F 316, must be viewed with the questions just stated.

5.6 THE DIAMETER RATIO: "LARGEST" PORE/"MEAN-FLOW" PORE

This section is directed to investigators employing the Coulter Porometer™ in examinations of filter media. This instrument (from Coulter Electronics, Hialeah, Florida) provides a convenient way of characterizing flat-sheet filter media, but the user must understand how to view the results.

The procedure for using the instrument is simple:

1. Cut out a small disk from the filter medium.
2. Wet it with a low-vapor-pressure oil, provided by Coulter, and of known surface tension.
3. Mount it in the instrument.
4. Press the "go" button.

The instruments proceeds to apply increasing air pressure to the disk while measuring the increasing flow of air and, while also storing the pressure-flow information. That is, the instrument traces Curve C of Figure 5.2.

After all pores have apparently been opened, the air pressure is decreased to trace Curve D of Figure 5.2, and the instrument stores that information. Then, on comparing the two curves, the instrument reports the size of the largest pore, and the size of the mean (median) flow pore.

Mayer (1993) provides data to show the limitations of the instrument and, apparently, the operator.

But before discussing Mayer's data, let's look again at Figure 5.2. There we see that the largest pores, corresponding to Point 1, have diameters of 2.8 units, whereas the median-flow pores, corresponding to Point M, have diameters of 0.65 units. The ratio is 2.8/0.65 = 4.3. And, as discussed in Section 5.3.3, if we call the bubble point Point 3, because our setup is not sensitive enough to measure lower air flow rates, then the ratio is only 1.47/0.65 = 2.26. That is to say, the ratio of the size of the "largest" pore to the median-sized pore decreases with the sensitivity with which we measure the bubble point. Now let's look at Mayer's data in Table 5.1.

Table 5.1
The Diameter of the "Largest" Pore, L, μm (from Bubble Point), and the Diameter of the "Mean Flow" Pore, MFP, in some Bag-Filter Media

L	119	50	18	14	4	3
MFP	13	6.1	3.6	3.6	2	2.8
L/MFP	9.15	8.19	5	3.9	2	1.07

Of interest here is that the L/MFP ratio increases with an increase in the size of the "mean-flow pore." Indeed, the diameter of the largest pores increases with the square of the diameter of the mean-flow pore.

Mayer agrees with the author (private communication) that

- in filter media with a random array of building blocks, the ratio L/MFP must be essentially equal from one filter medium to the next,
- the failure of the Porometer™ to report such consistent results lies with the small (not measured) air flow rate through the fine media when searching for the bubble point.

Thus, while we can believe the reported diameters of the median-sized pores in Table 5.1, we question the values of the "largest" pores, for at least, the fine-grade media.

This author understands that a newer Porometer™ has the wherewithal to adjust the sensitivity of the air flow meter. Perhaps an even newer model will report where, on Curve C of Figure 5.2, the bubble point is measured, to answer the question *"What fraction of the flow distribution is embraced? "*

5.7 DISTRIBUTION OF PORE VOLUME (RATHER THAN FLUID FLOW)

BACKGROUND

5.7.1 The Mercury-Intrusion Test

In this test we place a sample of porous material in a chamber, evacuate the air, and then slowly push in mercury, while keeping tract of the volume of mercury introduced, and of the increasing pressure to push it in. After it appears that all the voids are filled, we slowly reduce the pressure to allow mercury back out of the chamber. Mercury does not wet the walls of the pores, so it has to be forced in, and, once in, it wants to get out.

The kind of results obtained are shown by the generic plot in Figure 5.3, in which we see an *intrusion* curve and an *extrusion* curve. The term *hysteresis* refers to the fact that two different curves are obtained. Writers who address fluid intrusion in general explain that two different curves are obtained because of a likely difference in wetting angles—one angle in, the other out. But because mercury does not wet pore walls, there can be no consideration of the dry-wet history in this case.

Further, the two different curves are reproduced when mercury is again forced into the medium, then allowed to come out. In addition, the shapes of the curves do not change with a change in the *rate* at which pressures are increased or decreased.

The *intrusion* curve shows the distribution of the diameters of pores, or "throats." The *extrusion* curve shows the size distribution of larger chambers. The force at which the mercury expels itself is related to the diameter of the chambers, rather than to the throats, which remain flooded as the chambers empty (Conner et al. 1984; Coyne et al. 1986).

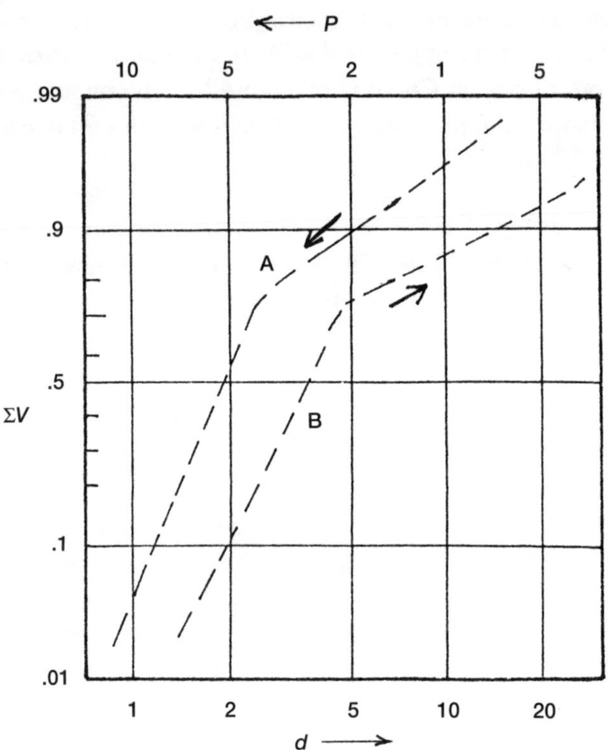

Figure 5.3. Example of results of measuring the pore-volume distribution in filter media via the mercury-intrusion test. The vertical scale represents the cumulative void spaces remaining as mercury, under a pressure, *P,* top, horizontal scale is increased to define Curve A, the *intrusion* curve. The lower horizontal scale indicates pore diameter, *d*, a function of 1/*P*. Once the void spaces are filled, and the pressure is relaxed, mercury comes out along Curve B, the *extrusion* curve, to define the diameters of chambers larger than the pores, or "throats." If the curves had defined straight lines on these coordinates, we would have seen log-normal distributions.

Obviously, the curves of Figure 5.3 do not show log-normal distributions, or are not close to one, as in Figure 4.2. The departure from an expected straight line on this plot can be caused by the following:

- There can be errors in normalizing the data (so as to plot values on the vertical, probability-scale). Many investigators show their plots on linear/linear scales, in which case it is difficult to determine the true end of the distribution and thus make the kind of plot shown in Figure 5.3
- In the experimental setup, mercury approaches the sample from all sides, rather than from one face to the other.
- Two different pore-size distributions may very well be present.
- In the case of a microporous plastic membrane, the force needed to fill the pores with mercury can also compress the membrane, thereby giving erroneous results.

Interestingly enough, Grace (1956), in examining a piece of woolen felt with the mercury-intrusion test, did find a true log-normal distribution of pore diameters. The geometric standard deviation was 2.0—that which we expect from the discussions in Chapter 4.

5.7.2 The Liquid-Drainage Test

Imagine this experiment: Saturate a filter medium with water. On top of the mixture apply a slowly increasing pressure of air, while measuring the increasing amount of water expelled, but don't apply so much air pressure that air flows through the medium.

This kind of test has been performed on beds of glass beads (Haring and Greenkorn 1970) and sand (Bear 1972). The resulting data define curves with the same shape as Curve 2 in Figure 4.3.

5.7.3 Problems with Reported Data

While many investigators have reported fluid-intrusion data for different types of filter media, much of the data is in the form of a variety of small, linear/linear plots with few tick marks on the graphs. From their presentations it is difficult to extract the detailed data necessary to make the log-normal-probability-type plots shown here. Which is to say, it is difficult to compare the shape of one writer's curve to that of another's to see if indeed the pore-size distributions follow the model in Figure 4.2, or in Figure 4.3, or another model.

We hope that if investigators, doing future fluid-intrusion studies, present a small graph, they will also present a table with enough data for the reader to make a cumulative plot on log-normal probability paper, such as in Figures 4.2, 4.3, and 5.3.

Or, instead of addressing *cumulative* data, the investigators will show a *differential* plot. But, *not* on a linear/linear scale as the present Figure 4.1. Instead, show the plot on log/log paper. In such a plot we see a dome -shaped curve, as shown in the lower part of Figure 6.4. From the shape of that curve we can extrapolate the ends to see the size of the largest and smallest pores, within whatever portion of the whole distribution we want to include.

6
Particles in Fluids
Meanings of Particle Sizes and Particle-Size Distributions

6.1 INTRODUCTION

BACKGROUND mixed with *INSTRUCTIONS*

Because the aim of filtration lies in separating particles from a fluid (liquid or gas), we employ a variety of measurements to judge the *clarity* of the fluids feeding the filter medium and emerging from it.

This chapter begins by addressing the various meanings of clarity.

One of the meanings is the particle-size distribution, which includes the meaning of the "size" of an individual particle.

Because writers measure and express both particle size, and the distributions of sizes, in different ways, it's important that the investigator know these ways and can choose either

- the way that puts the best face on filtration results, or
- the more truthful, analytic way.

6.2 CLARITY OF THE STREAMS

Clarity obviously addresses visual appearance. To put numbers to that appearance we employ a turbidity scale. We say that a fluid is clear when a turbidity value lies below a certain reading.

Where filtration is employed to sterilize a liquid, we look for live microbes in the filtrate. We place a sample of the filtrate aside at a cozy temperature to see if it becomes cloudy after a few days and (or) gives off a gas. Or, we place a sample in a petri dish and later count the numbers of colonies, each colony corresponding to one original microbe. One definition of sterility is the state of having less than one microbe per 100 ml. One definition of clarity is the state of no particles larger than size x per liter.

The *mass concentration* of solids in a filtrate is ordinarily determined by passing a sample of the filtrate through some very fine filter paper or membrane and then weighing the recovered solids. Some writers refer to this measurement as the "gravimetric level" of the particles. In some cases the recovered particles are so few and so small that our judgment of the particle concentration lies only in the degree to which the filter has been stained, if indeed the particles have a different color than the paper or the membrane.

Where "clear" water is to be fed a reverse-osmosis unit, the clarity of that water is measured by passing a sample through a standard-grade, microporous membrane to see how much water can be passed before the membrane plugs with accumulated solids (described in Chapter 12 as the Silt-Density Index).

Clarity can refer to the concentration of oil dispersed in water, or vice versa.

Clarity can have special meaning. For example, in the paint industry one desires a paint to be free of globs, but one does want the presence of small particles. Thus, clarity, in this case, means a low amount of globs, where *low* is defined.

When a liquid is proclaimed free of fibers, the analyst must tell us the smallest fibers (length and diameter) he or she is able to determine.

Or, to bring us to the subject to be discussed next, we examine the particle-size distribution in the fluid. But before doing that, we must understand the meaning of size. If we are to write a standard method of reporting particle-size distributions, we must first agree on the procedure for measuring and reporting size.

Indeed, as we will see in Chapter 7, when we compare the particle-size distribution in the feed stream to that in the filtrate, in order to reach a measure of filtration efficiency, we must agree on how we are going to make *that* comparison. Thus, in the two sections that follow, we first discuss methods of measuring particle size, and then discuss the different meanings of the particle-size distribution.

6.3 MEANINGS OF PARTICLE SIZE AND HOW MEASURED

With a spherical particle, the meaning of size, diameter, or volume, is straightforward. And if we know the density of the material, we can deduce the mass.

If the particle is a well-defined crystal, we can possibly agree on what we mean by size. Even if the particle has some irregular shape, we can also possibly agree on the meaning of size, as we discuss below.

Further, because of the different kinds of commercially available automatic particle counters, we are able to obtain enough counts, by size, so that we can realize a statistically significant number of counts, and thus reach a meaningful particle-size distribution.

Size can mean the longest end-to-end distance as we view the particle, under the microscope, lying heavy-side down. Indeed, instruments called *image analyzers* project the microscopic view of collected particles onto a TV-like screen, which is then scanned for particle counts in such a measurement of size. By viewing many portions of the area of an analytical-filter paper or membrane, onto which we have collected particles, we can obtain enough counts to reach a statistically significant count.

Size can mean *volume*. One type of automatic particle counter, the *electrical-resistance* type, counts particles suspended in an electrolyte by drawing the suspended solids through an orifice, on each side of which is an electrode. As a single particle moves through the orifice, the instrument senses an increase in resistance between the electrodes. The greater the resistance, the larger the volume of the particle. Obviously, the concentration of particles must be small enough so that two or more particles don't pass through at once. When they do, the instrument counts one large particle rather than two or more small ones, and suffers an error called *coincidence*.

Size can mean the *area of a shadow* cast by a particle as it passes under a light. Here, we employ an instrument called an *optical* counter. This instrument senses particles as they flow suspended in either a liquid or a gas. The less light passing from the emitter to the receiver, the larger the particle. And like the electrical-resistance counter (and, indeed, like the image analyzer), the optical counter can suffer the error of coincidence.

When we want to count particles over a broad range of sizes, we realize that, in most cases, the numbers of small particles are orders-of-magnitude greater than the numbers of large particles (as we see below). Thus, to avoid coincidence with the small particles, we present dilute concentrations to the particle counter. To obtain a good count of large particles, we present more concentrated samples to the counter. In either case, we present two or more samples to the counter, for each end of the spectrum, to enable us to obtain counts that are statistically sound.

That is to say, the meaning of the "largest particle" lies in statistical reasoning, as discussed in Chapter 5, where we addressed the meaning of the "largest pore" in the filter medium. And while the meaning of the "smallest particle" also lies in statistically reasoning, it also lies with the sensitivity with which the counter can discern small particles.

All three of the automatic, particle counters described above are calibrated with spherical particles of either glass or latex beads. Such spheres are available in separate batches, where, in each batch, all particles are essentially the same size; and many different sizes are available in separate batches.

6.4 PARTICLE-SIZE DISTRIBUTIONS

In counting different-sized particles in a distribution, the above instruments group the counts into narrow size ranges. For example, the electrical-resistance counter groups the measurements of volume into ranges within each of which the volume of the largest particle is twice that of the smallest. That is to say, in each range, the largest particle is $2^{1/3}$ = 1.26 times the "diameter" of the smallest particle. In the present discussion, we follow that procedure. Thus, we are able to view either the *numbers* of particles, or the *volumes* (masses) of particles versus the diameters.

A fourth type of instrument directly measures the mass distribution, by *sedimentation analysis.* With this instrument, a well-stirred suspension is placed in an X-ray beam, after which the particles begin to settle. The large particles settle faster than the small ones, and the instrument provides a continuous-line, and cumulative-mass, printout of the mass distribution, on a scale from zero to 1.0 (0 to 100%), versus the *Stokes diameter.*

From the data provided by all the above *counters,* we can also draw a chart showing the *cumulative* mass, or numbers, versus the diameter.

6.5 COMPARING DIFFERENT PARTICLE COUNTERS

Here's what an ASTM committee did in an attempt to reach a standard meaning of particle diameter. (The American Society of Testing and Materials, now call themselves only ASTM.) The committee provided different labs, each having different types of particle counters, with a sample of a specific lot of AC (Air Cleaner) Fine-Grade Test Dust (essentially silica, also called Arizona road dust). These labs were also provided with standard-sized latex beads for calibrating their instruments. Each lab was asked to report the cumulative-number distribution versus the particle diameter. The results of this venture are shown in Figure 6.1 (Johnston and Swanson 1986).

We see that, for the most part, all the labs found the same kind of distribution. That is, each of their curves could be superimposed over the others. The only disagreement, aside from the counts of small particles (discussed in section 6.6.3), is the meaning of particle diameter.

At the time, the NFPA (National Fluid Power Association) defined diameter as the longest end-to-end distance. Thus, it became obvious that an investigator, using a counter other than the image analyzer, would simply multiply his or her diameter values by a certain factor to arrive at the "standard" meaning of diameter.

The ASTM effort reflected in Figure 6.1 led to the writing of ASTM F 660.

(*Text continued on page 59*)

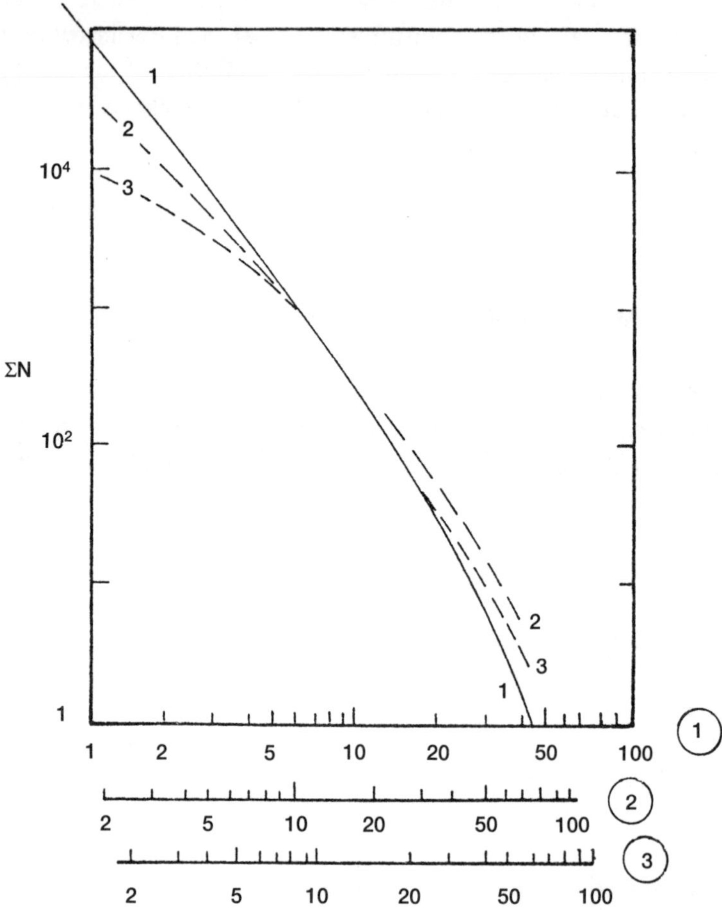

Figure 6.1. The particle-size distribution seen in AC, Fine-Grade Test Dust by three different kinds of automatic particle counters. The vertical scale shows accumulating numbers of particles with decreasing diameters, μm, on the horizontal scale, for a unit mass of test dust. Line 1, corresponding to horizontal Scale 1, is from the electrical-resistance type counters; Line 2, the image analyzer; and Line 3, the optical counter. Each instrument was calibrated with latex beads of certified diameters (Johnston and Swanson 1982).

(*Text continued from page 57*)

6.6. MEANINGS OF THE PARTICLE-SIZE DISTRIBUTION IN AC, FINE-GRADE TEST DUST

This test dust has been and continues to be used in laboratory filtration tests to characterize filter media according to procedures of ASTM, NFPA, and SAE (originally, the Society of Automotive Engineers). In the Figure-6.1 study, notice that investigators using the optical counters reported fewer small particles than did investigators using other instruments. At the time of that study it seemed apparent that workers using the optical counter had a prejudged view of the count of small particles, and, therefore, imposed that view into their report.

That is to say, the NFPA taught investigators using an optical counter to calibrate their instruments with AC, Fine-Grade Test Dust, specifying the distribution shown by Curve 3 of figure 6.1. In what follows, we look back on how the NFPA fell into that error, now corrected, and how, in hindsight, the group could have avoided it. This exercise in hindsight enriches our understanding of particle-size distributions.

6.6.1. Deducing the Number Distribution from the Mass Distribution

In the days before automatic particle counters were available (or readily used), the investigator's knowledge of the particle-size distribution in AC, Fine-Grade Test Dust was based solely on the brief data in Table 6.1, provided by the supplier, showing a differential *mass* distribution of particle sizes

Using the data of Table 6.1, we will now, in steps, construct a cumulative-*number* distribution so that we may compare it to the distributions in Figure 6.1.

With the data in Table 6.1, we first construct the graph of Figure 6.2 in an attempt to show the cumulative mass of particles versus diameter. Having drawn the best estimate of the curve in Figure 6.2, we then transfer it to log-normal graph paper shown in Figure 6.3. Because the curve in Figure 6.3 is a straight line—below a particle diameter of 35 μm—it becomes obvious that we have a truncated log-normal distribution, with a mass-median diameter of 8 μm, and a geometric standard deviation of 4.0.

Table 6.1

Particle-Size Distribution in AC, Fine-Grade Test Dust

Stokes diameter range, μm.	Mass Fraction, %
0–5	39 ± 2
5–10	18 ± 3
10–20	16 ± 3
20–40	18 ± 3
40–80	9 ± 3

The original supplier of this test dust, the AC Spark Plug Division of General Motors, no longer offers this material. It is now offered, in various grades (see Figure 6.6) by Powder Technology, Inc., Burnsville, Minnesota 55337, and is called SAE Test Dust.

Given these data, we construct a differential log-normal distribution of this type, showing it as the bottom, stair-step curve of Figure 6.4. That is, this bottom series of steps shows the relative masses of particles within the many narrow ranges of particle diameters.

To draw the top stair-step curve of Figure 6.4, representing the (relative) numbers of particles in each diameter range, we construct the height of each step from the bottom curve via the ratio (height of the bottom step)/(diameter cubed).

Finally, we construct the continuous, top (solid-line) curve by accumulating the stair-step values. The continuous curve represents the cumulative-number distribution of particle diameters that we compare to the curves in Figure 6.1. We see more small particles in this curve than are indicated by Curve 3 in Figure 6.1

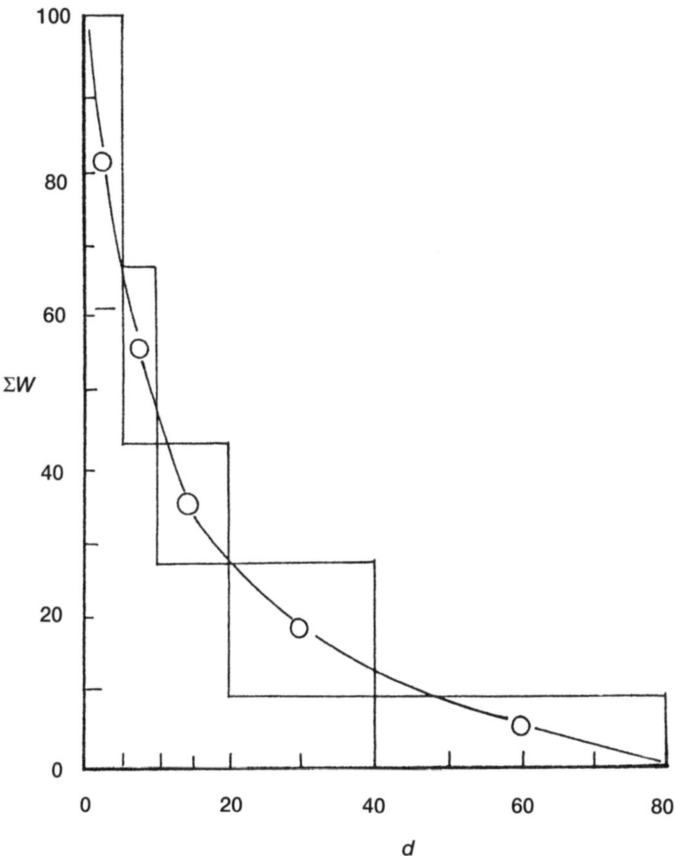

Figure 6.2. Plot of data in Table 6.1, showing the cumulative mass, ΣW, of particles with decreasing diameters, d (μm), in AC, Fine-Grade Test Dust.

6.6.2 Cole's Method Of Reaching The Number Distribution

We don't know if the late Fred Cole went through the exercise just described, but we do know that he painstakingly examined AC, Fine-Grade Test Dust under a microscope by counting the relative numbers of

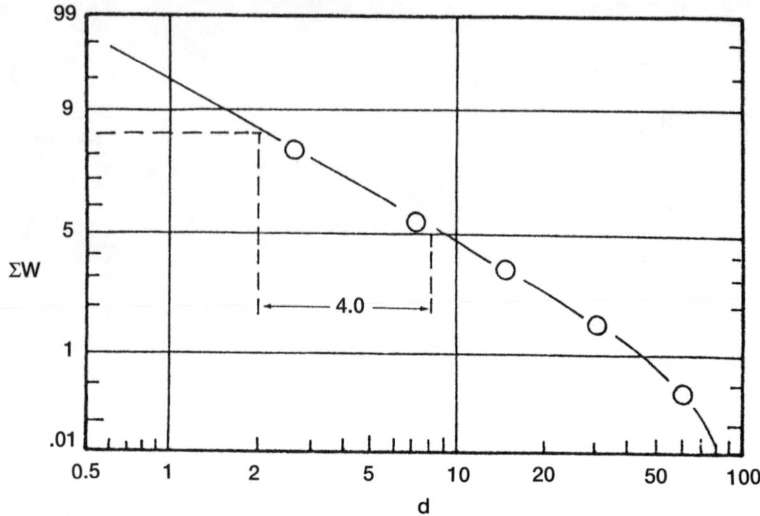

Figure 6.3. Plot of the curve in Figure 6.2 on log-normal probability paper. If not for the truncated, large-particle end of the spectrum, the distribution follows a log-normal distribution (because of the straight line) with a mass-median diameter of 8 μm, and a geometric standard deviation of 8/2 = 4.0.

particles in the diameter (longest end-to-end distance) range of 10 to 40 μm (Cole 1966).

He then reasoned that the distribution likely follows a log-normal distribution, which led him into making an interesting plot of particle counts. The kind of graph paper Cole used is shown in Figure 6.5. When he plotted his cumulative counts of particles versus the square of the logarithm of the particle diameter, *in his diameter range of 10 to 40* μm, he obtained a straight line.

6.6.3 An NFPA Standard

Cole's fellow members of an NFPA committee took his plot and ran with it. They extended the straight line, as shown in Figure 6.5, down to a particle diameter of 1 μm and up to 100 μm.

(*Text continued on page 64*)

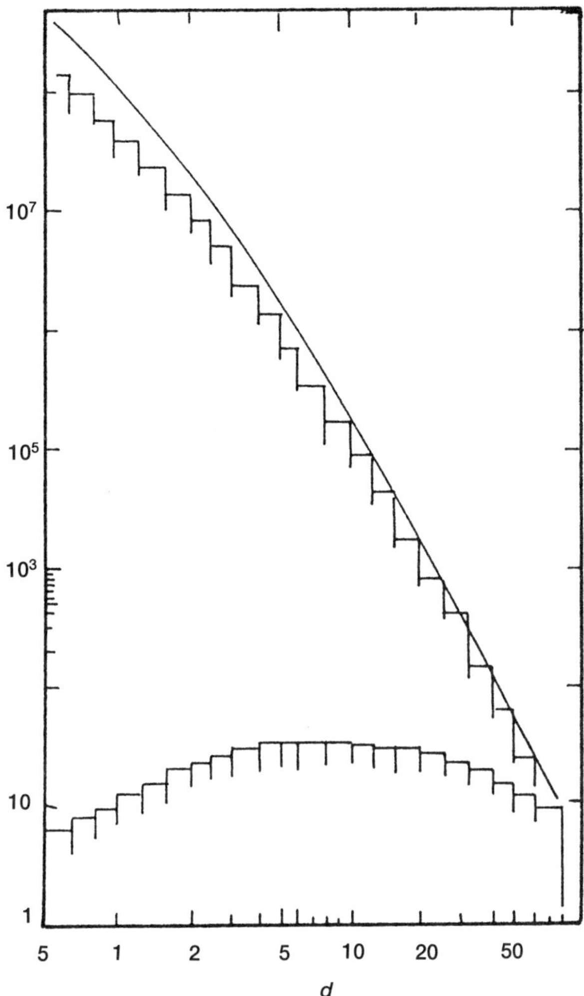

Figure 6.4. Alternative views of the particle-size distribution in Figure 6.3. The bottom stair-step curve shows the relative *masses* of particles, (vertical scale) in each range of particle diameters, *d.* The upper stair-step curve shows the corresponding *numbers* of particles in each diameter range. The upper continuous curve shows the cumulative numbers of particles, and is meant for comparison to Curve 1 of Figure 6.1.

(*Text Continued from page 62*)

From that line they produced a table to show the cumulative counts of particles in a unit mass of AC, Fine-Grade Test Dust versus particle diameter, in the diameter range of 1 to 100 μm (Fitch 1970).

That table became the standard definition of the particle-size distribution in AC, Fine-Grade Test Dust. Indeed (as mentioned previously) operators of optical particle counters were instructed to calibrate their instruments by way of the data in that table.

Thereafter, writers addressing particle-size distributions, displayed their data by means of the Figure-6.5 kind of plot rather than the kind of plot in Figure 6.1. Indeed, Bauman (d, pp. T-35 to T-39) shows that at least five separate standards-writing groups employ the graph paper of Figure 6.5—with the curves turning up for small particles—rather than the graph paper of Figure 6.1, which shows straight lines for small particles and which has the ability to reach particles smaller than 1 μm.

The straight line in Figure 6.5 corresponds to Line 3 in Figure 6.1. That is to say, the straight line in Figure 6.5 shows fewer numbers of small particles than are really there.

By 1974, investigators, using the electrical-resistance type of counter, were displaying particle-count data for AC, Fine-Grade Test Dust via Line 1 in Figure 6.1, instead of Line 3, which the NFPA made standard.

And by 1978 investigators were arguing that when a person "twists the dials" on an optical counter, so that the instrument will read the particle-size distribution as Line 3 of Figure 6.1, he or she will run into trouble when performing a filtration test. That is, on performing a filtration test with this dust, and on looking at the particle-size distribution in the filtrate (to be compared with the feed stream), the investigator would obtain misleading data concerning the efficiency with which a filter stops 1-to-10-μm particles (Johnston and Schmitz 1974; Johnston 1978).

But by 1981, the NFPA realized its error (Campbell and Iwanaga 1981), and it now instructs investigators to calibrate optical counters with latex beads.

(*Text continued on page 66*)

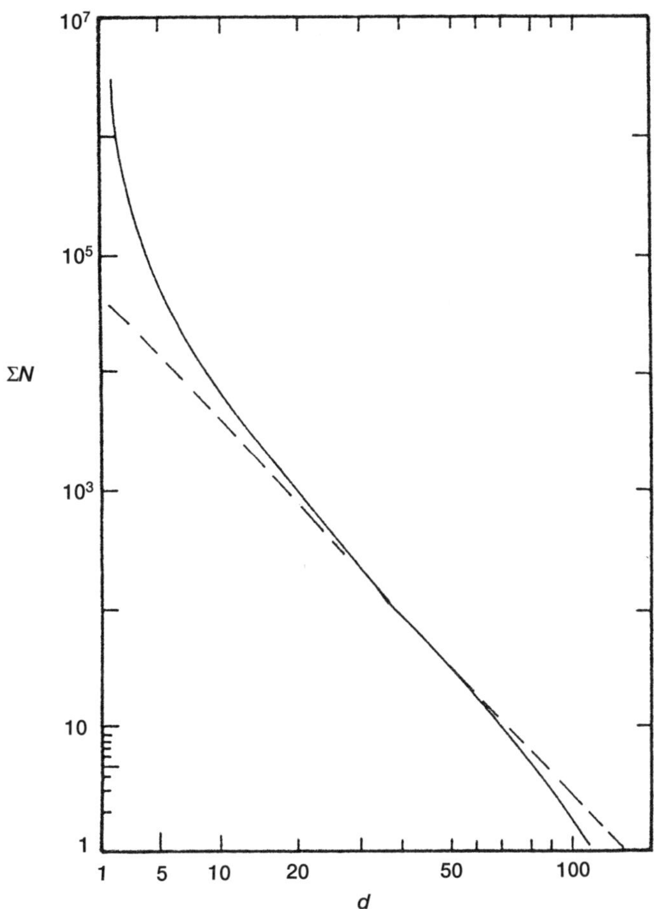

Figure 6.5. Plot of the particle-size distribution in AC, Fine-Grade Test Dust on Cole's (1966) chart paper. Cole, on determining the cumulative number of particles, ΣN, in the diameter range, d, of about 10 to 40 μm, saw a straight line. The NFPA extended the line (here the broken line) to include particles as small as 1 μm, and as large as 100 μm, and designated that to be the distribution, corresponding to Curve 3 of Figure 6.1. The solid-line curve here corresponds to Curve 1 in Figure 6.1, and to the continuous curve in Figure 6.4. The d scale is laid out as $(\log d)^2$. That is, *zero* on this scale corresponds to $d = 1$; *one* to d = 10; and *four* to $d = 100$.

(*Text continued from page 64*)

Apparently, the NFPA has not addressed the data in Figure 6.1, which shows that one instrument will report a particle size differing from that of another instrument, when both have been calibrated with latex beads. Yet, NFPA tests only employ the optical counter.

Thus, while an earlier, standard definition of particle size (for AC, Fine-Grade Test Dust) was the longest end-to-end distance, we see no present, standard definition, except that presented in ASTM F 660.

6.6.4 Other Grades of Siliceous Test Dusts

As stated in Table 6.1 (page 60) the AC Spark Plug Division of General Motors no longer offers Arizona road dust. The new supplier of this siliceous material is Powder Technology, Inc. (PTI). That material is now called SAE Test Dust.

PTI offers a description of the particle-size distributions in their products by way of an analysis with the electrical-resistance-type counter (the Coulter Counter™). But rather than describing number distributions, PTI reports normalized, cumulative-mass distributions. Further, in normalizing the data (that is, "percentizing" it) PTI assumes that the small-particle end of the distribution stops at 0.5-µm-diameter particles. That is, on accumulating the mass from the large- to the small particle end of the spectrum, PTI stops at 0.5 µm, describing the mass accumulated so far as all of the mass (100%).

PTI does this because SAE wants to see normalized, mass distributions. We plot PTI's data in Figure 6.6.

Having mentioned the three different kinds of particle *counters* (image analyzer, optical, electrical resistance) and the sedimentation device, we warn the analyst to be alert in using any of these instrument, or indeed *any other kind of instrument.*

The able analyst *will understand* how an instrument

- is calibrated
- measures particle *size*
- computes, and reports, particle-size distributions.

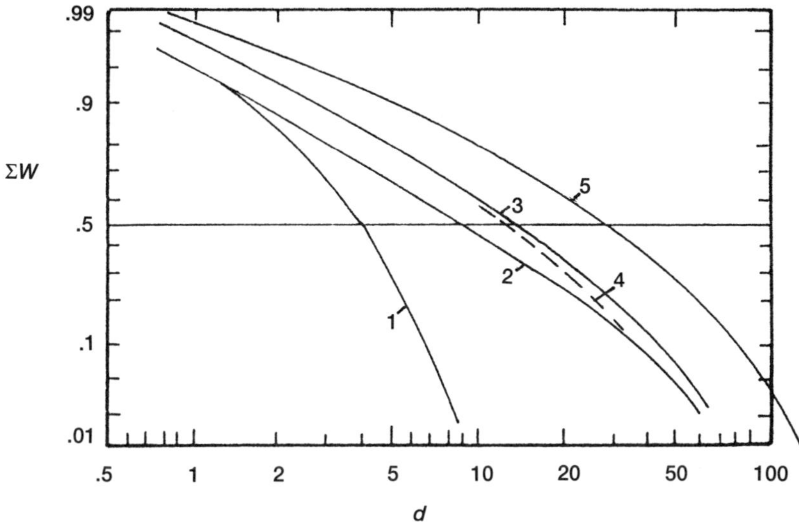

Figure 6.6. Particle-size distributions in different grades of SAE Test Dust, showing the cumulative mass, ΣW, of d-diameter (μm) particles, as determined with a Coulter Counter™, and assuming the absence of particles smaller than 0.5 μm (so the distributions can be "percentized"). Line 1 = Ultrafine Grade; Line 2 = Fine Grade; Line 3 = Medium Grade; Line 4 = "5 - 80"; Line 5 = Coarse Grade (Francis 1992).

6.7. MATHEMATICAL MODELS OF PARTICLE-SIZE DISTRIBUTIONS

The particle-size distribution in any powder, or in any suspension, depends, of course, on the origin of those solids, and if they have been classified by some technique of screening or sedimentation. Various math models have been proposed for describing particle-size distributions, since some distributions can be described by some math models.

Johnston (1976), on seeing the particle-size distribution depicted by Line 1 of Figure 6.1, suggests the expression

$$\Sigma N = A(L - x)x^{-c} \tag{6.1}$$

where: ΣN = the cumulative numbers of particles per unit volume of suspension, or per unit mass of powder.
A = a concentration index,
L = diameter of the "largest" particle. (Line 1 of Figure 6.1 falls toward zero at some large diameter.)
x = particle diameter of interest,
-c = the (negative) slope, on the log/log plot, of the straight line where x values are small.

Further, from Equation 6.1, he derives an expression for the cumulative volume of particles, ΣV, versus diameter, by way of differentiating Equation 6.1, and including a *shape factor*, ϕ

$$dN = A[-cLx^{-1-c} - (1-c)x^{-c}]dx$$

$$dV = \phi A[-cLx^{-2c} - (1-c)x^{3-c}]dx$$

$$\Sigma V = \phi A\left[\frac{cLx^{3-c}}{3-c} + \frac{(1-c)x^{4-c}}{4-c}\right]_x^L \tag{6.2}$$

Johnston also provides a graph showing solutions to Equation 6.2; and Heywood (1969) discusses shape factors, ϕ.

On replotting Line 1 of Figure 6.1, to show the horizontal scale as *volume* (instead of *diameter*), and constructing the horizontal scale to show three decades of volume to one of the present diameter scale, we find that we can essentially superimpose one curve over the other. To describe this second plot, Bader (1970) provides the expression

$$\Sigma N = K(V^{-n} - L^{-n}) \tag{6.3}$$

where: ΣN = cumulative number of particles,
K = concentration index,
V = volume of the particle size of interest,

-n = negative slope, on the log/log plot, of the straight line for small V values,
L = volume of the perceived largest particle.

Bader also provides an expression for the cumulative volume of particles

$$\Sigma V = \frac{\text{nK}}{\text{n -1}}\left(V^{1\text{-n}} - \text{L}^{1\text{-n}}\right) \tag{6.4}$$

where K, n, and L are the same as in Equation 6.3.

In those cases where we have "percentized" the volume distribution—that is, we know for sure the sizes of the largest and the smallest particles, and we have *accounted for all those sizes* so we can construct a plot like Figure 6.3—we can sometimes see the Rosin-Rammler expression describe that distribution.

$$\Sigma W = \exp{-\left(\frac{x}{x_r}\right)^k} \tag{6.5}$$

where: ΣW = cumulative mass fraction, accumulating from high to low diameters,
x = particle diameter
x_r = reference diameter
k = a factor addressing the breadth of the distribution.

By *reference diameter* is meant that when the ratio $x/x_r = 1.0$, the value of ΣW is 0.368.

Values of k less than 1.0 describe broad distributions, greater than 1.0, narrow distributions. For example: In gas-filtration tests, some test aerosols are employed where the geometric standard deviation of the mass distribution is as small as 1.3. Such a narrow distribution can be approximated with $k = 4$ (Johnston 1990).

7
Describing Filtration Efficiency

When I use a word, it means just what I
choose it to mean—neither more nor less.
—Humpty Dumpty, in *Through the Looking Glass,*
by Lewis Carroll

Turn on the blue light; the man wants a blue suit.
—An anonymous clothier

7.1. "WHEN I USE A WORD. . ."

INSTRUCTIONS

The literature of filtration is somewhat muddled. Much of that disorder lies in describing filtration efficiency.

Chemical engineers refer to filtration as a separation process. Particles are separated from a carrier fluid, or vice versa. Hence, the efficiency of this specific separation process is called filtration efficiency.

Many filtration writers speak of the *particle-removal characteristics* of a filter medium. Some, realizing that *characteristics* doesn't really say anything, speak simply of *removal.* Obviously their eyes are on the fluid rather than on the solids. Perhaps, other writers, with eyes on the solids, speak of the fluid-removal characteristics.

In gas filtration writers refer to *arrestance* (the filter medium incarcerates the particles).

It seems that many writers want to avoid the *E* word; hence, we see terms such as *purification coefficient, decontamination factor, titer reduction ratio, microbiological safety index, retention, rejection, penetration, sieving coefficient, filtration ratio,* and *Beta ratio.*

Each writer wants to coin his own word—and, it gets worse. All too often we see writing along this line: "Penetration is that percentage of the feed particles penetrating the filter medium."

> *Saponified coconut oil is coconut oil*
> *that has undergone saponification.*
> —Anonymous

To describe filtration efficiency we obviously compare the clarity, or cloudiness, of the feed stream to the filtrate. Yet, as we saw in Section 6.2, clarity can have different meanings, depending on how we measure it; and, of course, differing in what kind of measurements suit our needs.

When we measure the mass concentration of particles in the feed stream, C_1, and in the filtrate, C_2, filtration efficiency, E, is defined as

$$E = (C_1 - C_2)/C_1 \tag{7.1}$$

Alternatively, the C values can refer to turbidity measurements, or, in the paint industry, can refer to the concentration of globs.

Where a measure of clarity refers to the staining of an analytical filter paper, then the C values can refer to the degree of staining—where *degree* has been defined.

When a measure of clarity refers to the volume of liquid required to plug a given area of a standard microporous membrane, then *plug* has to be defined (see Silt-Density Index, Section 12.6.1).

In addressing filtration efficiency, it is mathematically more convenient to employ the term *filtration ratio*, R, or the reciprocal value *penetration*, *Pen*, (passing all the way through).

$$R = C_1/C_2\,; \quad Pen = 1/R \tag{7.2}$$

When values of R are very large, it is convenient to speak of *log R*, called the *log-reduction ratio*. For example, when $E = 0.999\ 999$, and $R = 10^6$, then $\log R = 6$.

During the course of a filtration run, changes occur in filtration efficiency—for a variety of reasons. And changing the conditions of the operation (for example, changing the velocity of the fluid) changes filtration efficiency.

Thus, whenever investigators report filtration efficiency they owe it to their readers to report exactly how they measure efficiency (and perhaps why they use that method). They are also obliged to report the conditions of the run along with the times in the run when they make such measurements.

7.2 WHEN CLARITY REFERS TO PARTICLE-SIZE DISTRIBUTIONS

Chapter 6 points out the lack of a standard meaning of particle diameter. As we see in Figure 6.1, one approach has been to simply calibrate a particle counter with latex beads, then use the particle diameters as reported—and then telling the reader what has been done.

But even there, different kinds of optical counters report different results (Verdegan et al. 1992). Further, latex beads do not stand up to oil—in using those beads to calibrate an optical counter that will analyze oil streams. Hence, one development under way for oil-filtration tests lies in providing a standard test dust provided in an oil suspension, where the particle-size distribution has been certified, and particle size refers to Stokes diameter. That suspension will be used to calibrate the counter (McBroom 1993).

In any event, investigators owe it to their readers to explain exactly how they measure particle "size." Further, investigators owe it to their readers to report their measurements of the particle-size distributions. such as in Figures 7.1 and 7.2, and not to report "normalized" distributions, such as in Figure 6.6, or as in Table 6.1, for three reasons:

1. In normalizing the data the investigator assumes knowledge of the distribution of all particle sizes, when in truth he or she is uncertain of the concentration of particles at both ends of the spectrum.
2. The reader may want to repeat the work, and one check method can be a look at the particle-size distribution in the feed stream, such as is depicted in Figures 7.1 and 7.2.
3. When the investigator looks at the *capacity* of a filter medium (the mass of test particles fed before the medium plugs), the reader must see the particle-size distribution in the fed stream. Remember, different particle-size distributions in the feed stream lead to different values for the particle-feeding capacity of a medium.

Even the very language of filtration begs clarification. Some examples:

The ASTM (STP 975, Vol. II, p. x) discourages the use of the adjective "particulate" as the plural noun, i.e., *particulates* (just as it is poor English to coin other nouns, such as *organics*, *schematics*).

The group further discourages the use of *contaminant* to mean particles (call a spade a spade); after all, some contaminants are soluble.

The NFPA has continued (since 1973 or before) to use *contaminant*, in referring to Arizona road dust (ANSI/NFPA T3.10.8 RI-1990).

The SAE is more specific, it refers to this test dust as *particulate contaminant* (SAE J1858 , 1988).

Does "throughput," mean *flow rate* or *volume filtered*?

As we see in Chapter 14, writers in cross-flow filtration employ muddled language.

And, on a somewhat similar Humpty Dumpty note, we beg writers to be more specific in referring to "wet" and "dry" filtration. Is there a "damp" and a "sloshy" filtration?

7.2.1 Comparing the Particle-Size Distributions in the Feed Stream to Those in the Filtrate

During the course of a test-filtration run, the investigator routinely measures the clarity of the filtrate and compares it to the clarity, or cloudiness, of the feed stream. What follows is an example of one such sample comparison, showing the different ways investigators compare particle-size distributions.

In this example the feed stream contains AC, Fine Grade Test Dust. Figure 7.1 shows the number distribution of that test dust, along with the number distribution in a filtrate. The continuous curves show the cumulative numbers of particles per unit volume of liquid. The stair steps show the numbers of particles within narrow diameter ranges.

Figure 7.2 shows the mass (or volume) distributions. The continuous curves show the cumulative masses. The stair steps show the masses within different diameter ranges.

NFPA procedures look only at the continuous curves of Figure 7.1, and compare them at particle diameters 5, 10, 15, 20, and 30 μm.

The NFPA coined the term *Beta ratio*, and uses it as follows: Consider 10-μm particles on the continuous curves of Figure 7.1. In each sample, we see the *number* concentrations of that diameter-*and-larger* particles in the feed stream and in the filtrate. On comparing those numbers, we obtain the Beta ratio of 14,000/3,100 = 4.5, or, more specifically written, $\beta 10 = 4.5$. The NFPA's test-report sheet does not mention filtration efficiency. It only reports the Beta ratio.

The SAE procedure follows the same method, but does not use the expression *Beta ratio.* Instead, it uses *filtration ratio* to mean Beta ratio, relates that to filtration efficiency, and provides a chart to draw, as shown in Figure 7.3.

ASTM methods, by contrast, look only at the numbers of particles in each of the narrow ranges of particle diameters. For example, in Figure 7.1 one compares the stair-step values for 10-to-12.6-μm particles to obtain the ASTM's filtration ratio of 7,040/2,340 = 3.0.

(*Text continued on page 79*)

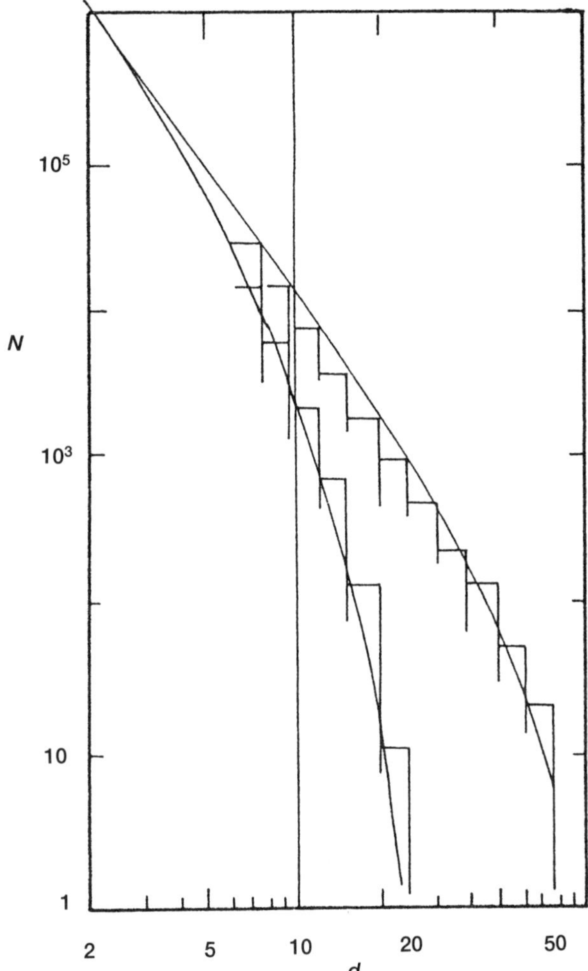

Figure 7.1. Examples of two ways of describing the particle-*number* distributions in streams around a filter medium. The smooth curves represent the *cumulative* numbers, N, of particles, per unit volume of a stream, for diameters d (μm). The stair-step curves show the numbers of particles *within* specific diameter ranges. The upper curves show the feed stream, the lower curves, the filtrate.

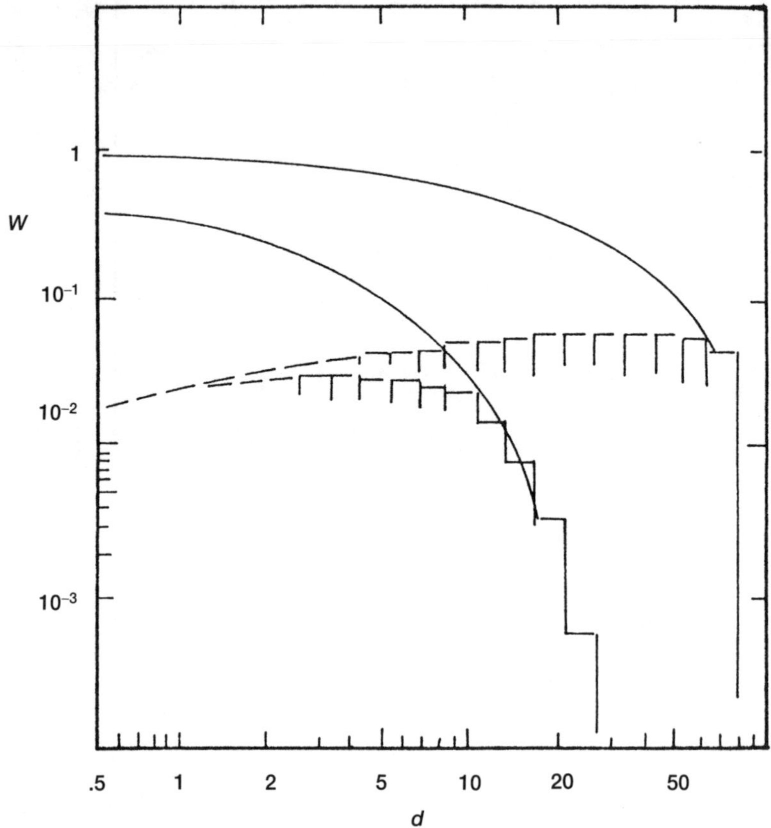

Figure 7.2. Two ways of describing the *mass* (or volume) distributions of particles in streams around a filter medium (from the *number* distributions of Figure 7.1). The smooth curves show the *cumulative* masses of particles, *W*, per unit volume of each stream, against particle diameter *d* (μm). The stair-step curves show the masses of particles within specific diameter ranges.

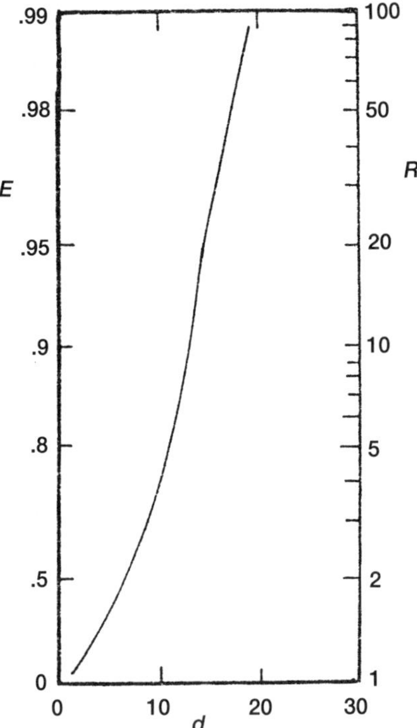

Figure 7.3. The SAE J1858 chart method of describing filtration efficiency, *E*, or filtration ratio, *R*, versus particle diameter *d* (μm), from the smooth-curve data of Figure 7.1. That is, in Figure 7.1, the *numbers* of 10-μm-*and-larger particles* in the feed stream are 14000 (per unit volume of that stream), and in the filtrate are 3100. The ratio, *R* = 1400/3100 = 4.5. In this chart the *R* scale is logarithmic; the *E* values corresponds to *R* values via Equations 7.1 and 7.2. The *d* scale is linear.

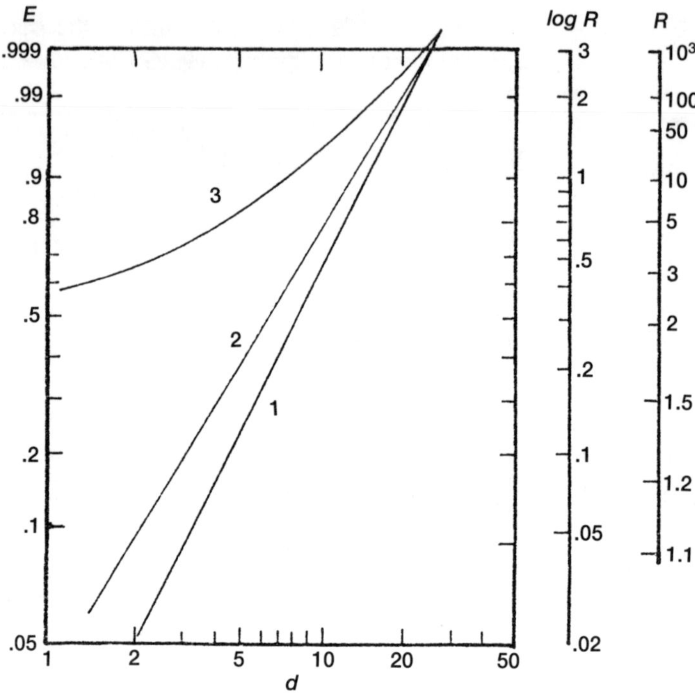

Figure 7.4. Curve 1 is the ASTM method of representing filtration efficiency, *E*, versus particle diameter, *d* (μm), from the stair-step curves in either Figure 7.1 or 7.2; e.g., the ratio of the masses, or numbers, of particles within the 10–12.6-μm-diameter range is the filtration ratio, *R*, of 3.0, corresponding to a filtration efficiency, *E*, of 0.67. By contrast, the SAE chart in Figure 7.3 defines the present Curve 2. Curve 3, sometimes used, compares the cumulative-mass curves of Figure 7.2. The present *E* scale is laid out from the *log R* scale (Equations 7.1 and 7.2), which itself is laid out on a log scale. That is, the straight line, Line 1, can be described by

$$R = \exp(ad^n), \text{ or } \log R = ad^n, \text{ or } \log \log R = \log a + n \log d,$$

where *a* describes an overall filtration efficiency, and *n* is the slope of the line, in this example 2.0.

(*Text continued from page 74*)

Alternatively, ASTM methods look at 10- μm particles (all by themselves) as follows: Notice the slopes of the continuous lines in Figure 7.1 at where d = 10 μm. They are –2.6 for the feed and –3.9 for the filtrate. The ratio of the slopes is 2.6/3.9 = 0.67. We now multiply the Beta ratio (the SAE's R value) by this slope ratio to obtain the ASTM's R value: 4.5 • 0.67 = 3.0.

ASTM methods (e.g., F 795, F 1170) further encourage investigators to plot results on the kind of chart paper shown in Figure 7.4. (The reason for using this kind of chart paper is explained in Chapter 8.) Line 1 corresponds to the method encouraged by ASTM. That is, for 10-μm particles, the filtration ratio, R, is 3.0, and the filtration efficiency, E, is 0.67. On the other hand, SAE J858 presents Curve 2, plotting it as Figure 7.3).

Two other SAE methods, J806 and J905, look not at the particle-size distribution but look only at the total mass concentration of particles (without use of a particle counter). From Figure 7.2, and from Curve 3 in Figure 7.4, we see that the total mass efficiency of filtration (for particles larger than 0.5, or 1.0 μm) is just under 0.60.

Regarding Curves 2 and 3 in Figure 7.4, if we change the particle-size distribution in the feed stream of Figure 7.1, we will change the shapes of Curves 2 and 3 in Figure 7.4; but we will not change the shape of Curve 1. Indeed, Campbell and Iwanaga (1981) point out that changing the particle-size distribution in the feed stream does indeed change Beta ratios (corresponding to Line 2 in Figure 7.4). They go on to provide a nomograph to be used in converting a Beta ratio obtained from a "nonstandard" particle-size distribution in the feed steam to the Beta ratio obtained if AC, Fine Grade Test Dust had been used.

When coarse-grade test dust is used in the feed stream, writers refer to *Alpha* ratios.

This author, employing a Coulter Counter™ to examine the particle-size distributions in water streams around filter media, where the feed streams contained AC Test Dust, has seen that the efficiency with which a filter separates particle diameters of 0.5–1.0 μm corresponds to the efficiency with which turbidity is separated.

8
Different Views of Filtration

INSTRUCTIONS

This chapter, as a preface to the next chapter, is directed to both the filter user and the manufacturer.

In designing or choosing a filtration test, we must view the big picture before getting down to details. All too often investigators take a narrow view, and, thereby, fail to see how better to design or choose a meaningful test.

Probably the most popular filtration-test procedure, and certainly one of the first with such detailed instructions, is the one written by the NFPA, for the needs of that special field of study. In the absence of other such detailed test procedures outside of that special field, writers have used the language and views of the NFPA in addressing and defining the general field of liquid filtration. That is, many investigators have carried the thinking and the language of the NFPA procedure into areas where different thinking and different language are required.

What follows is a review of the different fields of filtration, which will provide us with the various views in and requirements of those different fields.

8.1 VIEWS OF THE NFPA AND THE SAE

A manufacturer of hydraulic-power equipment, or of gasoline or diesel engines, approaches a producer of filter cartridges with requirements for a cartridge (or element) that will filter hydraulic fluid, lube oil, or fuel oil. The cartridge must handle a certain, constant volumetric flow rate, and fit into a certain housing. In feeding the cartridge a specific oil stream containing Arizona road dust, the driving pressure will be allowed to climb to a certain top limit, as the cartridge collects particles.

The cartridge manufacturer, after designing and testing the cartridge according to specific NFPA or SEA procedures, reports that he can supply a cartridge that demonstrates

- certain Beta ratios (see Section 7.2.1) at separating test dust from the oil, and
- that a certain mass of test dust can be fed to the cartridge before the driving pressure reaches the upper limit.

The manufacturer must design a cartridge that is both efficient and long-lived (not to mention inexpensive and rugged). The manufacturer is required to provide the results of a *multi-pass* filtration test wherein the filtrate returns to the feed tank, while, at the same time, fresh test dust is continuously added to the feed tank, in order to maintain a constant concentration of test particles in the feed stream.

In this case, *constant concentration* refers to the *mass* concentration of test dust. Of course, with time the feed stream becomes richer in the *numbers* of small particles, which are not separated with the efficiency of large particles, but which have relatively little mass.

While the NFPA and SAE test procedures fix the properties of the oil (temperature, viscosity, and "electrical" properties), the filter manufacturer knows that the lower the approach velocity of the oil through the filter medium, the greater the filtration efficiency. Thus, in his cartridge, he crowds in as much filter-surface area as he can—in the form a pleated-paper-type filter medium.

8.2 VIEWS IN THE CHEMICAL-PROCESS INDUSTRY

In that wide arena called the chemical process industry, the filter manufacturer is not so restricted in providing the best surface area of the filter medium. That is, for a given-sized stream to be filtered (with its content of solids), the filter manufacture will recommend either the best number of cartridges, or bags, or the best area of a filter cloth that will be used in a plate-and-frame filter or in a continuous-belt type of filter.

If the solids are the material to be recovered, the filter cloth must easily release those solids. In conducting a filtration test, the filter manufacturer and (or) the user will employ actual samples of the process stream.

The tests can be quite simple, such as a *single-pass,* constant-pressure (vacuum) test on a lab bench. Or they involve a more elaborate setup, with pumps and other equipment. What one looks for are

- *filtration efficiency* as a function of fluid velocity, at a specific temperature, or perhaps at different temperatures, to address summer versus. winter conditions;
- the *capacity* of the medium, meaning how soon it plugs with accumulated solids, or how soon a certain depth of filter cake is formed; and
- whether the medium can be easily *backwashed* and used again.

The filter manufacturer may have to recommend a two-stage scheme —one filter medium and then another. And perhaps (when recovering solids is not the aim), the first stage would involve the use of a filter aid, while the second stage would involve mere "polishing" of the liquid.

Indeed, the second stage may require a medium that stops microbes, such as yeast cells, diameters of 0.6 to 0.8 μm, or the small test microbe *Pseudomonas diminita*, 0.2 μm.

Alternatively, the "polishing" filter medium, as used in the electronics industry, may be required to stop very fine mineral particles, while at the same time not adding soluble material to the filtrate.

The filter medium may be required to clarify a gas stream (see Chapter 10) by separating very small particles, including microbes.

The filter medium may be a bed of sand—a very thick medium.

The filter medium may have to be heat resistant, such as metal fibers used to withstand molten polymers, or a heat-resistant cloth used to clarify stack gases.

8.3 VIEWS IN CROSS-FLOW FILTRATION

So far, we've talked about "dead end" filtration. In many liquid-filtration problems it is advantageous to employ cross-flow filtration. Here the feed stream runs over the face of the filter medium at high velocities, as one portion of the stream passes through the medium, and the other portion sweeps away solids that would otherwise collect on and plug the medium.

But even here the medium has a finite life, although sometimes it can be backwashed and used again. This scheme requires balancing what portion of the feed stream passes through the medium and what portion does not. That which does not is recirculated back to the source of the feed stream, or treated in another fashion.

When the filter medium is a very fine membrane, or a bundle of microporous tubes (hollow fibers), even soluble materials are separated from the liquid.

8.4 SEPARATING IMMISCIBLE FLUIDS

In some situations a liquid to be filtered contains droplets of another, immiscible liquid (water dispersed in oil, or vice versa). In other situations a *gas* to be filtered contains droplets of a liquid. Further, in all these cases, solid particles are also suspended in the fluid to be filtered, making the separation more difficult.

8.5 GENERAL POINTS

It is essential to keep the following points in mind when contemplating the design of, or the act of, performing a filtration test:

- While the use of specific *test* particles and a specific *test* fluid, at a specific velocity—often viewed as the specific driving pressure—will provide the investigator data to rate a filter medium, the use of the *real* fluid, with its *real* particles, will yield different test results.
- But, instead of employing an artificial test, we may more easily *rate* a filter medium via considerations of materials of construction, thickness, porosity, and permeability.

9
General Considerations In Liquid Filtration

This chapter is essentially *BACKGROUND*, to guide the reader in designing a filtration test, or to help her or him interpret the test results of another investigator. Yet, some *INSTRUCTIONS* are also provided, in Sections 9.4, 9.7, and 9.8.

When the liquid does not contain droplets of another, immiscible liquid, but only suspended particles to be separated, and if we do not necessarily want to recover the particles, the investigator should consider the following points.

9.1 GENERAL MECHANISMS OF LIQUID FILTRATION

As particles approach the face of a filter medium, most of those particles larger than the pores obviously don't enter. Those that do enter some of the few large pores do not get very far, because the pore channel chokes them off.

As a medium-sized particle does enter the face of the filter medium it may go deeply into the medium, or even pass through, yet by and large it will pull away from the laminar-flow stream as the stream turns a corner, so that the inertia of the particle carries it to a pore wall. Once on the wall it will stick, by van der Waals forces (unless it is "shocked" off by sudden surges in fluid flow, or vibration). Remember, in laminar flow the velocity of the liquid on the pore wall is nil, and the particles are not easily washed away.

On the other hand, small particles tend to stay in the laminar stream (remember, in pipe flow, particles concentrate in the center). Yet, small particles exhibit more Brownian motion than large particles. Thus, there is the statistical chance that some of these small particles, in their random movements, will find themselves up against the pore wall. This Brownian movement increases with an increase in temperature, and (independently) with a decrease in the viscosity of the carrier fluid.

Further, the small particles may be attracted to the pore wall by differences in the zeta potential of the particle and the wall (more on this in Section 9.2).

And the lower the velocity of the carrier fluid, or the thicker the filter medium, the more time the particles have to randomly break out of the laminar stream and find the wall of a pore.

Thus, given a certain filter medium, we can more efficiently clarify a mobile liquid than a viscous one. And, as the reader has obviously concluded, we can more easily clarify a gas than a liquid, because of the extra-low viscosity of the gas.

9.2 THE ZETA POTENTIAL

At a solid-liquid interface, or a liquid-liquid interface, a collection of positive and negative charges is grouped onto the surface of the solid, or on the suspended liquid droplet. This "double layer" is made up of ions, in aqueous solution, that are held firmly to the surface, and in a more diffuse, mobile layer extending into the solution.

The resultant, net charge of the diffuse layer is equal in magnitude, but of an opposite sign, to that of the firmly held, or fixed, layer. Because of these electrical charges, there exists a difference in potential between the fixed layers and the bulk of the solution. This has been called the electrokinetic potential, or given the noncommittal name zeta potential, represented by ζ.

Thus, if the zeta potential of a particle in suspension is of an opposite sign to the zeta potential of the pore walls in a filter medium, the particles will be attracted to the wall.

Here, it is convenient to recall Cohn's Law relating zeta potentials to dielectric constants. For example, water has a dielectric constant, 70, greater than cellulose fibers, 5. Hence, by Cohn's Law, cellulose fibers, in water, have a negative zeta potential. Quartz (silica) has a dielectric constant, 4, also less than water, and hence has a negative zeta potential in water.

Jaisinghani and Verdegan (1982) provide a discussion of how we measure the zeta potential of a filter medium.

9.2.1 Examples of the Zeta Potential

From Cohn's law we deduce that in the filtering of silica test dust from water by means of a bed of cellulose fibers, silica particles are not attracted to the fibers. But when we precoat the fibers with, for example, a polyamine, we change the sign of the zeta potential. The result is that a filter medium so treated shows a very dramatic increase in the efficiency with which it captures silica particles (e.g., Cuno's Zeta Plus™ filter media).

That is to say, a coarse medium with such a coating can be made to filter as a fine medium. Indeed, this coating does not change the permeability of the medium. (For a given flow of water through the medium, the addition of the coating does not mean we must apply a greater driving pressure to obtain the same flow without the coating.) But we must be aware of the following:

- In employing a fine-grade medium without the coating, and in making an extended filtration run, we notice that filtration efficiency, as a rule, increases as the pores become clogged with the captured particles. We, thus, stop the run when the permeability drops to an unacceptably low level.
- However, in employing a coarse medium with the coating, thus allowing the medium to filter more efficiently, we notice that before we see a drop in permeability, filtration efficiency drops.

The reason for this latter action lies in the fact that the particles attracted to the pore walls have now "neutralized" the zeta potential of the walls; hence, filtration efficiency drops as other particles come along. We don't have the drop-in-permeability signal to tell us when to stop.

This is not to say we should avoid the advantage of fibers so treated. Indeed, where we employ a fine medium without the treatment, and require a certain driving pressure to obtain a certain flow rate, switching to a coarse medium with the treatment enables us to enjoy the same filtration efficiency but with as little as one tenth the driving pressure. This advantage is useful when our liquid is very viscous. But we must be careful in knowing when to stop the run. We must know, from test beforehand, the "electrical" capacity of the medium to clarify the specific liquid.

9.3 "SIEVING" FILTRATION

Putting aside any impact of the zeta potentials on filtration efficiency, along with any chance that a particle smaller than the pore may find the pore wall, one goal in filtration is simply to employ a filter medium with pores small enough to stop the particle sizes of interest. Put in another way: It doesn't matter that the length of the average-sized pore is a hundred or a thousand times the average diameter, we just want to employ a filter medium with a small flow-average pore diameter. Or, more stringently, we want to employ a filter medium in which the "largest" pores are small enough. That way we will stop all particle sizes of interest, regardless of the viscosity and the velocity of the liquid.

To address this subject statistically, we begin by recalling the most probable flow-pore-diameter distribution discussed in Chapter 4. In Figure 4.2 we plotted the flow-pore distribution of two separate filter media, one with a porosity of 0.35, and the other with a porosity of 0.75. In each case, the smallest pore diameter is 1.0 unit.

We now replot those curves on the chart paper in Figure 9.1. Notice that this is the same chart paper used in Figure 7.4. What we see in Figure 9.1 is that the curves—almost straight lines—are parallel. That is, if we consider that the smallest pore in the 0.35-porosity, left medium has the same dimension as the smallest pore in the 0.75-porosity, right medium,

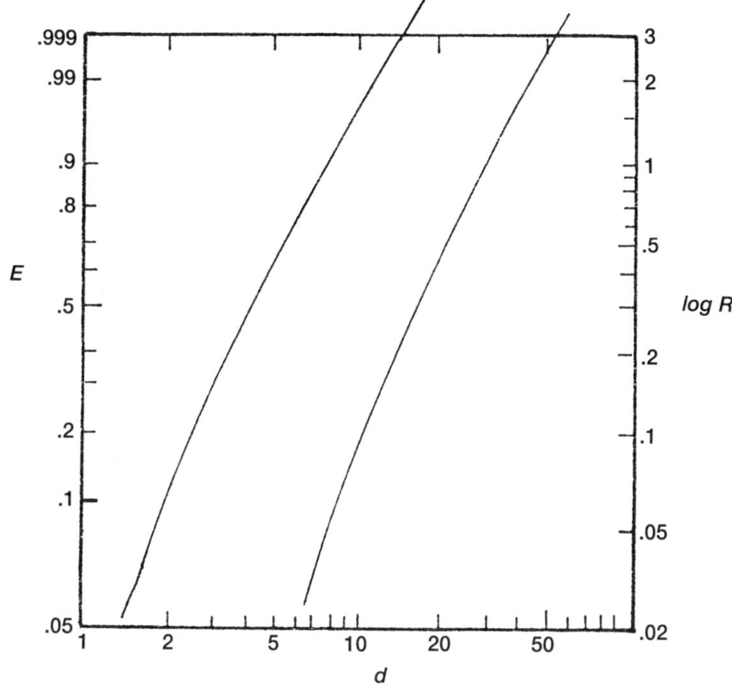

Figure 9.1. Chart paper of the kind in Figure 7.4 (page 78), onto which has been plotted the theoretical, flow-pore distributions of Figure 4.1(page 33). The horizontal axis represents pore diameter, with the smallest pores having a diameter of 1.0 unit. The left curve represents a filter medium of porosity, ε, of 0.35, the right is for one with ε = 0.75. Both curves are not too far removed from a straight line with a slope (measured against the *log R* axis) of 2.0—implying that the slope of Line 1 in Figure 7.4, also 2.0, represents sieving filtration.

then both curves tell us the following (looking at the right-hand curve as if the left hand curve were superimposed on top of it):

Half of the flow through the medium passes through pores with diameters larger than 16 units. Thus, in pure, sieving filtration (at the start of the run before the pores in the filter medium have been changed from collecting many particles), we expect this medium to stop half of the 16-unit-diameter particles; the other half will pass through the larger pores.

We also see that only 0.01 (1%) of the liquid passes through pores larger than 40 units. Thus, in pure sieving filtration we expect the medium to stop 40-unit-sized particles with an efficiency of 0.99 (99%), while the other 1% passes through the larger pores.

Thus, the left-hand vertical scale indicates the efficiency with which individual-sized particles, on the horizontal scale, are separated from the liquid, in sieving filtration. Note that the slope of the curve corresponds to Line 1 in Figure 7.4. This means that in the filtration example of Section 7.2.1, we saw the symptom of sieving filtration; the slope (*log log R* versus *log d*) was 2.0. Note further that where filtration is not pure sieving, but includes "absorptive" effects, we will see the lines of Figure 9.1 with less slope (see the examples in Figure 9.3).

But, whether the mechanism of filtration is pure sieving or it is highly absorptive, filtration efficiency changes with time as the medium collects more and more particles. Sometimes filtration efficiency increases; sometimes it decreases.

At this point we remind the reader that while we have the tool (Equation 1.7) to deduce the flow-average pore diameter from permeability data and thus can deduce the *pore-size* distribution, along with the diameter of the "largest" pores, and while we have the instruments to measure the *particle-size* distribution in the feed stream, we must not forget that we cannot simply match pore sizes to particles sizes in our effort to obtain sieving filtration. The methods of measuring pore size differ from those measuring particle size. Indeed, what do we mean by the actual or absolute result of either kind of measurement?

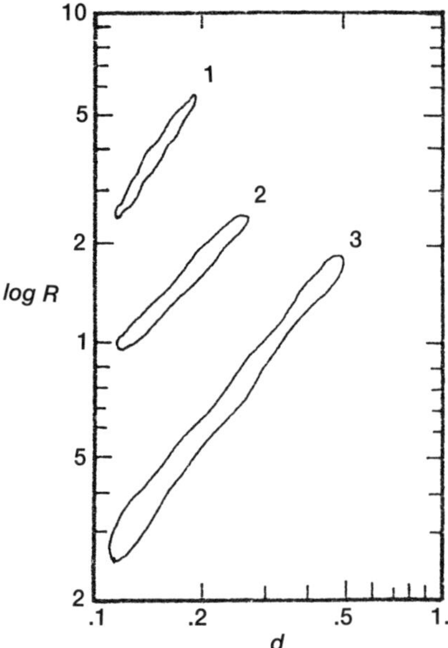

Figure 9.2. Data reported by Grant and Zahka (1990) in feeding water-borne suspensions of latex spheres, of which diameters *d,* range from 0.13 to 0.45 μm, to three different membranes, *rated* as Cloud 1, 0.1 μm; Cloud 2, 0.2 μm; and Cloud 3, 0.45 μm. Because the slopes are near 2.0, we deduce from Figure 9.1, that sieving filtration occured—a conclusion also drawn by Grant and Zahka, but from different reasoning.

9.3.1 Examples of Sieving Filtration

Grant (1988) and Grant and Zahka (1990) provide the data for our discussion in this section. Over the course of many experiments, Grant fed distilled-water suspensions of up to twelve different-sized spheres (with diameters: 0.12–0.55 μm) made of different materials (polystyrene latex, carboxylate-modified latex, polyvinyl toluene, and styrene butadiene) to three differently rated microporous membranes, 0.1, 0.2, and 0.45 μm. His results, plotted in Figure 9.2 show the symptom of sieving filtration (described in the previous section); that is, the slopes of his clouds are near 2.0. But Grant observed sieving filtration on the basis of a different model than that in Figures 4.2 and 9.1.

This author makes no attempt to explain Grant's complicated model; however, some idea of the model is obtained through the following excerpt of his writings (brackets and italics added):

> The model predicts the [filtration ratio of individual-sized particles] from [a] the initial reduced dimensionless filter coefficient, [b] the flow redistribution parameter, [and, c] the number of pore volumes of fluid passed through the filter media. . . . The *flow rate* through each pore is proportional to the *volume of the pores*. . . . [Point b] characterizes the way in which the flow paths through the media are redistributed as pores within the media become clogged with particles.

(Grant, as does many writers, refers to a single porous material, or bed, as a filter *media,* rather than a filter *medium.* Indeed, many writers refer to two or more materials as filter *medias.*)

Grant's model predicts that with continued filtration (the feeding of more particles) the filtration ratio falls, as particles "break through." For example, an initial *log R* value drops from 6 to 3 after 0.005 "pore volumes of particles have been removed," after which the *log R* values level off. The phrase in quotes apparently refers to the fact that after the membrane has been fed a certain volume of particles (that volume corresponding to 0.005 of the volume of pores), filtration reaches a more or less steady-state condition (within the time studied) as reflected in the present Figure 9.2.

Grant implies that during this time, the very small pores become clogged with collected particles, after which the flow stream is directed to larger pores, which allow particles through that would have been otherwise stopped by the small pores. (That is, no "neutralizing" of zeta potentials occurs, as discussed in Section 9.2.1.)

In other examples, Sueoka and Malchesky (1983) employed ultrafilter membranes to stop Angstrom-sized molecules of dextran and blood components. They plotted filtration efficiency (rejection) on a probability scale versus molecule size on a log scale. That is, they used the same chart paper as is presented in Figure 4.2. They obtained straight lines, indicating a true log-normal distributions of pore diameters with geometric standard deviations ranging from 1.51 to 2.13. Thus, if we assume they obtained sieving filtration, as we assumed above with Grant's experiments, then their results point to what we think is the pore-size distribution in filter media with a random array of building blocks.

9.4 RULES OF THUMB

INSTRUCTIONS

Grant's data reinforce two rules of thumb Johnston has proposed (1990, p. 57) regardless of the mechanism of filtration. These rules are directed to the filter user who is trying to find the best medium for his or her filtration problem, and who contemplates the use of different grades of filter media. We explain these rules with the following example:

In this example we have the ready choice of different filter media, all them with the same thickness, porosity, and material of construction. They differ only in the flow-average pore diameter, **d** (of Equation 1.7).

We (perhaps randomly) choose one to use in performing a filtration test. In this test we uses our own liquid of interest, and we decide what liquid-approach velocity to employ (staying in the viscous-flow region (see Section 1.1).

We find that our particle size of interest (or perhaps the turbidity) is separated with an efficiency of 50%. That is, $log\ R = 0.3$ (see Figure 9.1). But suppose we want is a medium that separates those particles with an efficiency of 95%. That is, we want $log\ R = 1$.

We have two choices:

- a medium with a smaller flow-average pore diameter,
- extra thicknesses of the present medium.

We keep open the choice of a lower fluid velocity, because we have no rule of thumb to apply in that situation.

Choice 1

To deduce what finer medium to use we employ the ratios

$$\frac{\log R_2}{\log R_1} = \frac{1.3}{0.3} = \frac{\mathbf{d}_1^2}{\mathbf{d}_2^2}$$

to find that $\dfrac{\mathbf{d}_1}{\mathbf{d}_2} = \dfrac{1}{0.48}$

That is, in choosing our second medium, with a flow-average pore diameter of $\mathbf{d}_2$, we will select and test the one with nearly half the flow-average pore diameter as the first medium, $\mathbf{d}_1$.

Choice 2

Alternatively, to decide what extra thickness, z, of the first medium to use we employ the ratio

$$\frac{\log R_2}{\log R_1} = \frac{z_2}{z_1} = \frac{1.3}{0.3} = \frac{4.33}{1}$$

Thus, we would employ four thicknesses of the first medium to obtain the desired filtration efficiency.

Whichever plan we choose to follow, we must increase the fluid-driving pressure by a factor of four to maintain the same fluid-approach velocity (Equation 1.7).

9.5 ABSOLUTE FILTRATION

New Kid in the filtration lab reads a test report, written by Ole Hand, in which the continuous lines of Figure 7.1 are displayed. The report says that 30-μm particles are removed with an efficiency of 100%. "You can't say that," says New Kid.

"Do you see any 30-μm particles in the filtrate?" answers Ole Hand.

"That's the point. You should have said the filtration efficiency is simply greater than 0.995; you don't know how much greater."

"Look at this bulletin from Clyde's Filters. It says a Beta ratio greater than 50 is absolute."

It was now quitting time. New Kid went for a drink.

An implied definition of *absolute filtration* is provided by ASTM D 3862, and D 3863. These test procedures employ a 47-mm-diameter disk of a microporous membrane, with a surface area of 10 cm^2. A 100-ml broth, containing 10^8 freshly grown test microbes, is placed on top of the membrane, then sucked through into a flask below (the vacuum level is not specified), after which the filtrate is examined for the count of live microbes.

When this filtrate is found to be sterile, the membrane is considered "absolute" at stopping microbes of that size. That is, if the microbe is *Serratia marcescens,* the membrane is *rated* as 0.45 μm. If the microbe is *Pseudomonas diminuta,* the membrane is *rated* as 0.2 μm.

Of course, *absolute* means a log-reduction value of "greater than 8." When a single microbe is found in the filtrate, the log reduction value is indeed 8. The filtration efficiency is 0.999 999 99 (eight 9s). As an earlier frame of reference, recall that the definition of heat sterilization lies with a log-reduction value of 6, which is a kill efficiency of 0.999 999.

Some investigators, attempting to demonstrate that a membrane is indeed absolute at stopping certain sized microbes, challenge a cartridge containing 4,500 square centimeters of membrane surface with a broth containing 10^{12} microbes ($2.2 \cdot 10^8$ microbes per square centimeter).

Where they find one telltale microbe in the filtrate, they proclaim that the cartridge failed the challenge test. Yet, even if they had found 200 microbes in the filtrate, they still would have demonstrated a log reduction value greater than 7 per square centimeter, *the original premise for "absolute" filtration.*

As a practical matter, membranes used in the pharmaceutical industry are never challenged with such enormous numbers of microbes. Indeed some investigators write that a 0.45-µm-rated membrane can "quantitatively" stop 0.2-µm-diameter microbes when the surface of the membrane is fed, say, only 10^3 microbes per square centimeter. Thus, some absolutes are more absolute than others. (See also Section 12.6.)

9.6 FILTRATION EFFICIENCY AS A FUNCTION OF TEMPERATURE AND APPROACH VELOCITY.

Figure 9.3 illustrates how changes in temperature and approach velocity change the efficiency with which waterborne particles of black iron oxide, or silica, are stopped by a filter paper. These tests are not prolonged tests, in which we look for drops in the permeability of the filter paper. The results simply show what happens before the paper becomes clogged with particles and is no longer the original filter medium.

Notice that Line C, representing iron oxide, is *steeper* than Line A, representing silica—at equal temperatures and velocity. Based on the premise illustrated in Figure 9.1 (that the slope reflects sieving filtration), we deduce that the filter paper more readily absorbs silica than iron oxide. Particle diameters are those determined with the Coulter Counter™, previously calibrated with latex beads (Johnston 1990).

Thus, in any standard filtration test, it is important to standardize the test particles, the temperature, the liquid velocity, and, of course, the liquid.

9.7 DEDUCING PORE SIZES FROM FILTRATION TESTS

INSTRUCTIONS

Don't do it.

Consider two filter media, each composed of the same material(s), each with the same porosity and permeability, but with one being thicker than the other. In this case the thicker material will be the more efficient filter.

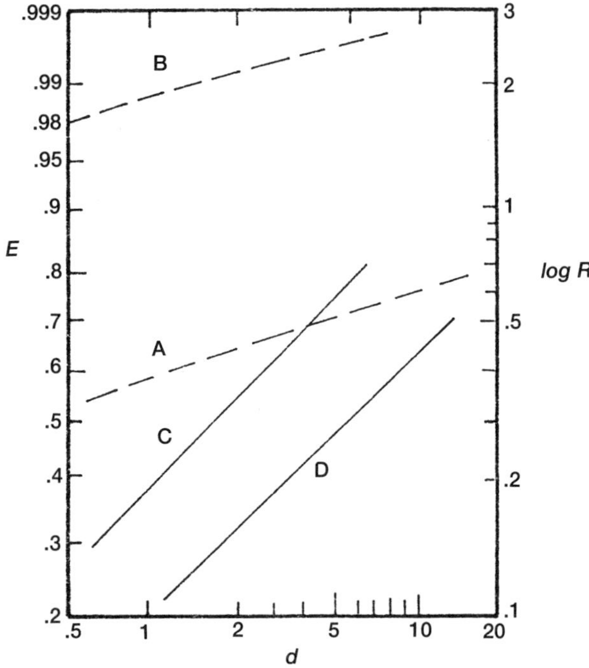

Figure 9.3. Efficiencies, *E*, with which a filter paper, separated particles of diameter, *d*-μm, from water (Johnston 1990).

Line	Particles	Temp., °C	Fluid-Approach Velocity, 10^4 m/s
A	AC Fine	21	7
B	Test Dust	77	7
C	Black iron	21	7
D	oxide	21	35

How, then, are we to deduce the pore sizes, even if we see the symptoms of sieving filtration?

Possibly, an investigator might measure filtration efficiency for many different thicknesses; then make a plot, for different particle sizes, of *log R* versus thickness (Rule of Thumb Choice 2, in Section 9.5); and then extrapolate the data down to some "minimum thickness" for values of *log R*. But even then, he or she must include studies of such variables as fluid velocity and viscosity, kinds of fluid, kinds of test particles, and temperature.

Recall Grant's data in Figure 9.2: A membrane that has been *rated* as have a maximum pore diameter 0.45 μm because it "absolutely" stopped microbes of that diameter (*Serratia marcescens*) allows latex spheres of the same diameter to pass through. Further, Johnston (1975) showed that such a membrane was only 90–98% efficient at stopping 0.45-μm-diameter particles of silica or black iron oxide.

9.8 REACHING A STANDARD RATING OF A FILTER MEDIUM

INSTRUCTIONS

By now we hope the reader will understand that there is no such thing as a universal-standard, liquid-filtration test. To be sure, many different "standard tests" do exist, with each one trying to apply a rating for a specific application. Indeed, many filtration tests apply a *pore-size rating* — based on a filtration test—which (in light of Section 9.7) is trivial, or as one speaker has put it, "like sounding brass and a tinkling cymbal."

The trouble is, most users of filter media have grown accustomed to such pore-size, or particle-size, ratings. Thus, the filter manufacturer is forced to tack this meaningless label onto his product.

For example, Alderete (1991), in showing the results of filtration tests among different cartridges, explains *nominal*, *absolute*, and *Beta* ratings. But he also points out that these tests, under specific conditions (Arizona road dust in water, at a specific temperature and flow rate), will yield different results under other conditions.

Many filter users, however, without the time to consider the many ramifications of filtration, want to hear only the rating (recall the quotes at the beginning of Chapter 7).

But at least the producers of membranes, who have been forced to rate their products by pore size rather than by the specific microbe tested, provide more information about their products than do producers of other filter media. And membrane producers provide useful information, reporting:

1. material(s) of construction,
2. thickness,
3. porosity,
4. data on fluid flowrate versus pressure drop, and
5. whether the material is homogeneous or one side is denser than the other.

Because a rating, based on a single "standard" filtration test, is trivial, the truly informative investigator addresses all these five points when describing and evaluating a filter medium.

This author recalls a meeting where the speaker, describing a resin-impregnated filter paper, reported such information as *basis weight* and *caliper*, but when asked about *porosity* essentially answered "What has that got to do with anything?"

Of course, *if* we knew the density of the fibers with their resin coating we could deduce porosity from the basis weight (mass per area) and the caliper (thickness). Or in the case of a non-woven cloth, e.g., of polyester, we could also deduce porosity from basis weight and caliber, knowing the density of polyester.

Manufacturers of filter paper and non-woven cloths will do a favor to filtration customers if they report porosity straight away, along with thickness, and *permeability* (Chapters 1 and 2).

Some writers and speakers confuse *porosity* with *pore size*.

10
Considerations in Gas Filtration

This chapter, like the chapter on liquid filtration, is a mixture of *BACKGROUND* and *INSTRUCTIONS.* Section 10.2 reviews the various test methods of gas filtration from which the reader may choose the one suited for his or her interest.

10.1 GAS FILTRATION VERSUS LIQUID FILTRATION

Gas filtration differs from liquid filtration in four respects:

1. In gas filtration, particle diameters in the range of 0.1 to 0.5 μm are separated with the least efficiency; that is, both larger and smaller particles are stopped with greater efficiency, as illustrated in Figure 10.1.
2. Apparently no one has seen this (or predicted this) in liquid filtration. Indeed, particle-sensing instruments for examining liquids are not available for detecting particle diameters smaller than about 0.5 μm.

3. A mathematical model of gas filtration successfully predicts the efficiency with which a single fiber held in a gas stream collects particles (ASTM STP 975, Vol. 1, pp. 1–74). The model considers the three mechanism of particle capture mentioned in Section 9.1 (direct interception, inertia, and Brownian diffusion). It includes the diameter of the fiber, the diameters of the particles, and the velocity of the gas stream (Pierce et al. 1990). Indeed the velocity is critical; a velocity that is too low or too high negatively affects particle capture (Jaroszczyk and Wake 1991). In addition, the model can include electrical aspects. (Trottier and Brown 1990).

   It then follows that the overall efficiency with which a fibrous bed collects particles becomes a function of the number of fibers in the bed, or, more specifically, the total area of the fiber surfaces. Indeed, the concept of pore size doesn't mean a thing in gas filtration. After all, the so-called pores are much larger than the particles stopped. The only pore model that can be applied is the one directed at that singular medium the track-etched membrane, with the equal diameter, straight-through tunnels (ASTM STP 975, Vol. I, pp. 74–94). The performance of the other types of membranes, produced by the solvent-cast method, better fit the fiber model.
4. Thus, with these gas-filtration models at hand—some even for granular filter media—we may more directly deduce the best filter medium for a given gas stream. Not so in liquid filtration, where we must make test runs to select the best medium.

However, gas filtration is similar to liquid filtration in at least three ways:

1. We may electrically enhance a gas filter. We may also include a resin with the fibers. The resin, a nonconductor, readily acquires a static charge, to capture particles. Or we may pass the gas-feed stream though a screen electrode just before the filter medium while on the downstream side of the filter medium another electrode is in place. A high voltage applied across the electrodes greatly enhances the filtration efficiency (ASTM STP 975, Vol. I, pp. 316–331)
2. A gas filter medium has a finite capacity. In gas filtration, writers

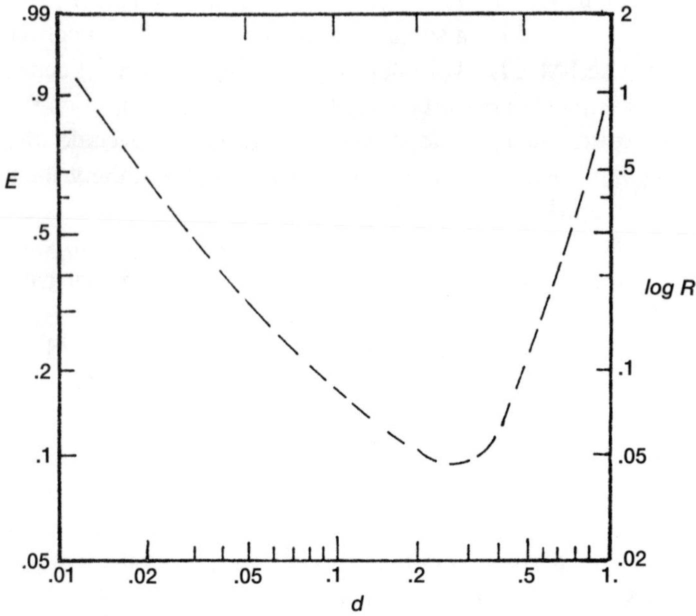

Figure 10.1. Typical results in gas filtration. Particles in the diameter, *d*, range of 0.1 to 0.5 μm or the most difficult to separate. Hence, test procedures in gas filtration employ test particles of such sizes. *E* = filtration efficiency, *R* = filtration ratio, defined in Equation 7.2, page 72. The *log R* scale, which is laid out on a log scale, defines the *E* scale, as in Figure 7.4, page 78.

call it *effectiveness*—the mass of particles collected per unit area before a significant drop in permeability (not to confused with *arrestance,* referring to filtration efficiency. Often, as in a baghouse, designed to clarify various exhaust gasses, the medium, the bags, can be shaken free of particles or back-blasted, and used again for many numbers of cycles. (ASTM subcommittee F 21.3 is currently addressing a test procedure involving a back-blasting or pulse-test procedure.)

3. Writers express filtration efficiency in all of the three different ways illustrated by Figure 7.4.

10.2 TEST METHODS IN GAS FILTRATION

As one might imagine, the test stand for gas filtration—usually air filtration—is more complicated than is a test stand for liquid filtration. The air must be kept at a certain temperature and humidity; it must approach the filter medium evenly, at a specific velocity; and the test particles must be suspended in the air stream as homogeneously as possible.

The test stand gets more complicated when we consider what test particles to use, how to generate them, how to suspend them in the air stream, how to count them, and how to compare the particle-size distribution in the feed stream to that in the filtrate.

Often we don't look at the particle-size distribution, we just consider the total mass of particles in the two streams. For example, Arizona road dust has been used where a known mass of dust is fed the medium, after which the medium is weighed to determine the increase in weight. Here we learn what mass fraction of the feed-stream dust has been *arrested* by the filter.

Jaroszcyk (1987) describes a device for feeding test dust.

Alternatively the dust is premixed with carbon black and lint, and during the run, samples of both air streams are passed through an analytical filter to compare the degrees of staining.

Powdered alumina (with a mass-median particle diameter of 5.2 μm) has been used as the test particles.

Dyes such as methylene blue (with a mass-median diameter of 0.5 μm) and uranin (with a mass-median diameter of 0.2 μm) have been used as test aerosols. Samples of the air streams are passed through an analytical filter, which is then examined for the degree of coloring.

Low-vapor-pressure oils have been converted into aerosols. Pierce et al. (1990) measured the efficiency (up to 0.999 99) with which a medium stops individual droplet diameters ranging from 0.07 to 0.3 μm. The air streams were examined with a condensation nucleus counter.

Latex spheres, freed from aqueous suspension and dispersed in air by means of a special device, have been used as test aerosols. Here the air streams are examined by an optical particle counter (ASTM F 1215).

Particles of NaCl have been used in diameters ranging from 0.001 to 1.0 μm. The sizes are measured with a differential mobility particle sizer. In this case the air streams are examined with a laser diode detector or a flame ionization detector (Simpson and Iverson 1989).

Of course, the size of the test particles to use depends on the size of the particles that one wants to demonstrate the filter will arrest.

More details on the above kinds of test procedures are provided in ASTM STP 965, Vol. 1, pp. 97–420. See also Jaroszczyk and Ptak (1985); Jaroszczyk et al. (1987), SAE Paper 872268; Edward Johnson et al. (1990); Brian Johnson et al. (no date); Remiarz et al. (no date).

A special kind of air filter is used in the pharmaceutical industry. Called a vent filter, it comes in the form of a pleated membrane cartridge, placed over a tank. The filter allows air in or out of the tank as it is filled with liquid or drained, but excludes the passage of microbes. The membrane, with a pretested bubble point (see Chapter 5) high enough to stop microbes, and is made of a hydrophobic material, yet is also kept hot with a steam jacket to prevent condensation of water vapor within the pores.

Another air filter used in the pharmaceutical industry is a glass-fiber mat designed to filter air fed to fermenters. This filter is pretested with an aerosol of corn oil (Meltzer 1987).

Dickenson (1992) provides many illustrations of gas-filtration media and systems.

11
Capacity Of A Filter Medium in Constant-Flow Filtration Use Of Filter Aids

11.1 PLOTTING PRESSURE RISE WITH RUNNING TIME

INSTRUCTIONS

In the situation where a filter medium is fed a stream at a constant flow rate—so that suspended solids in the stream (sometimes of known concentration) are fed at a constant rate—we must constantly increase the driving pressure to maintain that flow. Usually we employ a pump or fan large enough to provide the necessary increase in driving pressure while maintaining the constant flow rate.

During such a run, addressed in ASTM F 795, we make a plot of increasing driving pressure versus time, the latter corresponding to the volume of the filtrate or to the mass of particles fed the filter medium.

When the driving pressure reaches some upper limit, set by the conditions and apparatus for the operation (Mugg and Nicolazzo 1989), we stop the run. We learn what mass of those specific particles can be fed to

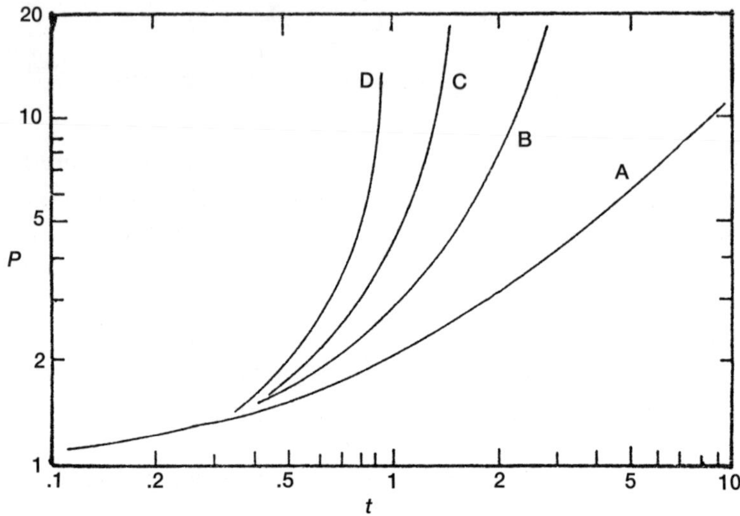

Figure 11.1. Math-model plots on constant-flow filtration, showing the required increase in driving pressure, *P*, with time, *t*. Curve A reaches a slope of 1.0.

Curve	Model Name	Equation
A	Cake filtration	11.1b
B	Intermediate blocking	11.3b
C	Standard blocking	11.4b
D	Complete blocking	11.5b

a unit area of the filter medium that is consistent with the fluid flow rate and the desired efficiency of filtration.

In doing this, we are measuring the capacity of the filter medium in relation to the specific conditions for the operation. One of these conditions is the position of the filter medium; that is, whether the medium is horizontal and fluid flows down on it or up to it, or whether the medium is vertical. Sedimentation also plays an important role, when the particles are large enough to settle.

In those cases where the collected solids are to be recovered, the limit may be set by the thickness of the filter cake (rather than by the driving pressure).

11.2 THE RATE OF PRESSURE RISE

It is also of interest to look at the *rate* at which the pressure rises. To observe this rate we make the above-mentioned plot on log/log paper. It doesn't matter what units of pressure or time we employ in the plot. The plot will show, for example, the factor by which the pressure increases with every, say, 10% increase in time.

We then compare our plot to the curves in Figure 11.1 by superimposing our plot over Figure 11.1 (making sure that the decade lengths of the log scales are identical). In doing this we can determine

- if the experimental curve rises very fast, as does Curve D of Figure 11.1,
- if the curve resembles Curve A the slope of which never climbs above 1.0,
- if the curve lies somewhere in between.

If the experimental curve resembles Curve A, we assume that

- the pores in the filter medium have not plugged with accumulated solids, and
- the collected solids, with the increasing thickness of the filter cake, have not compressed with increased pressure.

If the curve looks like Curve B, it is likely that, again, the pores in the filter medium have not been plugged, but that the cake of collected solids has compressed with increased pressure.

If the curve looks like Curve C or D, or some other steep curve, we may have a plugging of the filter medium and we may also have a filter cake that easily compresses.

In those cases where a substantial filter cake is formed, as when the solids are to be recovered, the data often plot as a straight line—or close to it, after, as seen in Curve A of Figure 11.1, the pressure across the cake becomes three or four times the pressure across the medium. But the slope is somewhat greater than the Curve-A slope of 1.0.

11.3 REMEDIES FOR FAST PRESSURE RISE

When the curve rises fast, similar to Curves C and D in Figure 11.1, we *may*, on employing a filter medium with smaller pores, obtain curves similar to Curves A and B. But the finer filter medium may not necessarily show an increase in capacity. Remember, capacity is determined by an upper limit of pressure. On switching to a finer medium we must provide a higher pressure at the start for the same fluid velocity. But at least it's worth trying a finer medium.

In liquid filtration, if the solids are not to be recovered, we can add a filter aid to the feed stream as we show by the example in Section 11.6. A filter aid consisting of granules or fibers larger than the particles in the feed stream gathers on the surface of the filter medium as a permeable cake while collecting the feed-stream particles.

11.4. MATHEMATICAL MODELS OF PRESSURE RISE

BACKGROUND

The math models cited in the caption to Figure 11 are derived from Grace's (1956) list of mathematical expressions that are derived from empirical data reported by Hermans and Bredée (1936).

11.4.1 Cake Filtration

At the start of filtration, before any particles collect, the filter medium exhibits a certain resistance to flow so that to achieve the desired (constant) flow rate we must apply an initial driving pressure, P_o, which, for the present discussion, does not include the driving pressure to overcome the resistance of the *housing*.

Then, with the increased deposition of particles, the required pressure increases to the point, as illustrated by Curve A in Figure 11.1, in which further increases are directly proportional to time, or volume of filtrate, V.

Grace writes

$$\frac{P}{P_o} = 1 + K_A Q^2 t \qquad (11.1a)$$

where K_A = a constant;
Q = flow rate, constant dV/dt
t = cumulative time.

At the start, where $P/P_o = 1$, t is essentially zero. To construct the Curve-A plot, in Figure 11.1, we let $K_A Q^2 = 1$ and make the plot from

$$P = 1 + t \qquad (11.1b)$$

Thus, in constant-flow filtration, when we construct a plot on log/log paper of pressure versus time and if the curve can be superimposed over Curve A in Figure 11.1, we readily see the symptom of ideal cake filtration (and we may want to go out and dance in the streets).

To actually deduce the rate constant, K_A, in Equation 11.1a, we make a linear/linear plot of P/P_o versus t. The slope is $K_A Q^2$.

Equation 11.1a is one of four expression derived from the general equation

$$\frac{dP}{dV} = KP^n \qquad (11.2)$$

In Equation 11.1a, $n = 0$. We now consider three other values for n.

11.4.2 Intermediate Blocking, *n* = 1.

Here, Grace's expression is

$$\ln\left(\frac{P}{P_o}\right) = K_B Q t \qquad (11.3a)$$

that we normalize for our generic plot of Curve B in Figure 11.1 to read

$$\ln P = t, \text{ or } P = \exp(t) \tag{11.3b}$$

To deduce the rate constant, K_B, we make a linear/linear plot of $\ln(P/P_o)$ versus t. The slope is $K_B Q$.

11.4.3 Standard Blocking, *n* = 3/2

Here Grace shows the expression

$$\left(\frac{P}{P_o}\right)^{-0.5} = \left(1 - \frac{QK_C t}{2}\right)^{-1} \tag{11.4a}$$

that we normalize for Curve C, in Figure 11.1 as

$$P = (1 - t/2)^{-2} \tag{11.4b}$$

To determine the rate constant, K_C, we consider Grace's alternative expression

$$\left(\frac{P_o}{P}\right)^{0.5} = 1 - \left(\frac{K_C}{2}\right)V$$

that guides us in making a linear/linear plot of $\left(\frac{P_o}{P}\right)^{0.5}$ versus V, from which we deduce $K_C/2$ from the slope.

11.4.4 Complete Blocking, *n* = 2

Here Grace's expression is

$$\frac{P}{P_o} = (1 - K_D t)^{-1} \tag{11.5a}$$

that we normalize for the plot of Curve D in Figure 11.1 as

$$P = (1 - t)^{-1} \tag{11.5b}$$

To deduce the rate constant, we consider

$$\frac{P_o}{P} = 1 - K_D t$$

where, on a linear/linear plot of $\frac{P_o}{P}$ versus t, the slope is K_D.

11.5 RESISTANCE OF THE FILTER CAKE

BACKGROUND

As discussed in this and the next two chapters, it is of interest to examine the resistance of the filter cake. This subject is commercially important when the cake is to be recovered.

In approaching this subject we make the following assumptions, stated more thoroughly by Willis and Tosun (1980):

1. The filter medium itself does not plug with solids. Or, if it plugs, the degree is negligible; the forming cake of collected solids immediately becomes the new filter medium.
2. In a Figure-11.1 kind of plot, the absence of Curve A, the symptom of cake filtration, does not necessarily mean the original filter medium has plugged. Rather, it can mean that the cake of collected solids, with its own permeability, has compressed with increased pressure to become less permeable. The resulting curve will have a final slope greater than the slope of Curve A.

Thus, on assuming the above as the true scenario, we proceed with the following reasoning.

11.5.1 The Meaning of Cake Resistance

The equation we now address, showing the combined pressure across the filter medium and the cake is (Tiller 1975; Tiller et al. 1977)

$$P = \eta u^2 t R_C + \eta u R_M \tag{11.6}$$

Letting $\mathbf{F}$ = force, $\mathbf{L}$ = length, $\boldsymbol{\theta}$ = time, and, in what follows, $\mathbf{M}$ = mass:

P = total pressure drop, $\mathbf{F/L^2}$,
η = viscosity of the filtrate, $\mathbf{F\theta/L^2}$,
u = fluid-approach velocity, $\mathbf{L/\theta}$,
t = time, $\boldsymbol{\theta}$,
R_C = resistance of the filter cake, $1/\mathbf{L}^2$,
R_M = resistance of the filter medium, $1/\mathbf{L}$,

The resistance of the cake is expressed differently than the resistance of the medium, with its fixed thickness, because we relate the resistance of the cake to the increasing *mass* of solids.

$$R_C = c\alpha_{av} \tag{11.7}$$

where, c = mass of solids per volume of filtrate, $\mathbf{M/L^3}$,
α_{av} = average specific resistance of the cake, $\mathbf{L/M}$,
The density, and resistance, of the cake is high next to the filter medium and low upstream from it.

Since we assume the resistance of the filter medium, R_M, remains constant, and the pressure drop through it is $P_o = \eta u R_M$, we write Equation 11.6 as

$$P - P_o = \eta u^2 t c\alpha_{av} \tag{11.8}$$

We now address α_{av} by way of an empirical constant, a, which is specific for the solids at hand, and by way of n, the compressibility factor of those solids. (This n is not the n of Equation 11.2.)

$$\alpha_{av} = a(1 - n)P^n \tag{11.9}$$

For non-compressible solids, e.g., sand, n = zero. For highly compressible solids, e.g., metal hydroxides, n approaches 1.0.

On assuming the pressure drop across the cake has reached the point where it is three or four times greater than that across the filter medium—see Curves A and B in Figure 11.1—we neglect P_o in Equation 11.8. Further we substitute Equation 11.9 into Equation 11.8 and thereby obtain

$$P^{(1-n)} = (1 - n)(a\eta c u^2)\, t \tag{11.10}$$

Equation 11.10 describes the pressure drop across the cake as a function of time; all other terms are constant. A log/log plot of pressure versus time shows a straight line (or close to one) with a slope of n + 1. This equation is employed in the following example.

11.6 EXAMPLE OF CONSTANT-FLOW FILTRATION WITH A FILTER AID

INSTRUCTIONS

Walton (1978, 1981), in a laboratory-scale operation, studied the use of a diatomite filter aid in clarifying beer. On first laying down a precoat on the filter medium, he also added this filter aid to the feed stream, calling that added filter aid *body feed,* and he fed the filter assembly at a constant flow rate.

In different experiments he employed different concentrations of body feed. In Figure 1 of his 1981 paper he shows linear/linear plots of pressure versus time. The present Figure 11.2 shows those data on log/log paper, from which we see the following:

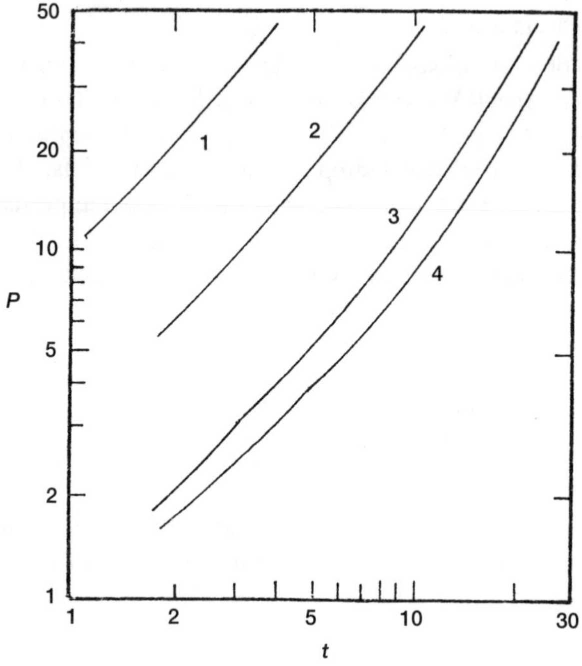

Figure 11.2. Walton's (1978) results in using a diatomite filter aid to clarify beer, in constant-flow filtration. P = driving pressure, psi; t = time, hours. With only a precoat on the filter medium, Curve 1, the filtration cycle lasted 4 hours. On also including diatomite in the feed stream, at three, increasing concentrations, Curves 2–4, the cycle time was increased to almost 30 hours. The diatomite concentration in the feed stream is expressed as the ratio of the mass of diatomite to the mass of suspended solids in the beer, and is called the body/feed ratio. This ratio in Curve 2 was 2/1; in Curve 3, 3/1; and in Curve 4, 4/1.

1. The experiment without body feed (Curve 1) that used only the precoat resulted in a pressure-increase rate faster than that for ideal cake filtration (Curve A of Figure 11.1) but not as fast as intermediate blocking (Curve B).
2. The use of body feed (Curves 2–4) did not change this rate, but it did extend the length of the cycle, meaning the volume of beer filtered (and the capacity of the filter medium).

3. With more body feed there comes the realization that only so much is useful.

The mass concentration of solids in the beer was determined by passing a sample of the beer through a 0.45-μm-rated, microporous, cellulose-ester membrane, then weighing the residue to deduce "the gravimetric level." Those solids, yeast cells and organic debris, were present at a mass concentration of 140 parts per million.

The precoat level consisted of 0.15 pound of diatomite per square foot of filter area.

In analyzing Walton's data, we employ Equation 11.10, which states that a log/log plot of P versus t should be a straight line, from which we can deduce the compressibility factor, n, and the constant a associated with the solids at hand.

What follows is an abbreviated way of analyzing data. Since Walton does not report the viscosity of the filtrate, η, we will look at the term $a\eta cu^2$ in Equation 11.10 as one unit.

Consider Curve 2 in Figure 11.2. We consider it a straight line, and see the slope to be 1.20, thus $n = 0.20$.

At the lower end of the line, $P = 6.0$ psi, when $t = 2.0$ hrs.

Employing our abbreviated use of Equation 11.10, i.e.,

$$P^{(1-n)} = (1-n) \cdot a\eta cu^2 \cdot t$$

we write

$$6.0^{0.80} = 4.19 = 0.80 \cdot a\eta cu^2 \cdot 2.0$$

to find that $a\eta cu^2 = 2.62$.

When $t = 10$ hrs, the upper end of the line, we calculate

$$P^{0.80} = 0.80 \cdot 2.62 \cdot 10 = 20.95$$

so that $P = 20.95^{1/0.80} = 44.8$ psi.

What this abbreviated analysis is meant to convey is that once we deduce the value of $a\eta cu^2$ and we know the value of all items in that term except a, we deduce the value of a, after which we deduce the value of α_{av} via Equation 11.9.

Two final comments about constant-rate filtration:

1. There exists an *optimum* rate for obtaining the most volume of filtrate and collecting the most mass of solids during a cycle. Of course, the lower the rate, the longer the cycle (Purchase 1977).
2. The time required for constant-rate filtration is twice that of constant-pressure filtration to reach the same volume of filtrate and the same pressure (Purchase 1977; Tiller et al. 1977).

12
Capacity Of A Filter Medium In Constant-Pressure Filtration

12.1 PLOTTING FILTRATE VOLUME VERSUS TIME AND FLOW RATE VERSUS TIME

INSTRUCTIONS

In the chemical-process industry, liquids are generally moved around at constant flow rates, and the previous chapter addressed constant-flow filtration.

In the laboratory it is more convenient to study filtration under conditions of constant pressure. In studying cake filtration we make separate runs at separate pressures to learn the compressibility of the cake.

In a lab setup for constant-pressure filtration, liquid is pushed through (or sucked through) a filter medium as we gather data concerning the cumulative volume of filtrate and (or) the drop in flow rate versus time (ASTM F 796). We end the test when the flow rate drops to, say, one tenth the starting value.

For whatever driving pressure we study, we make one of two separate plots, on log/log paper, of volume of filtrate versus time or of flow rate

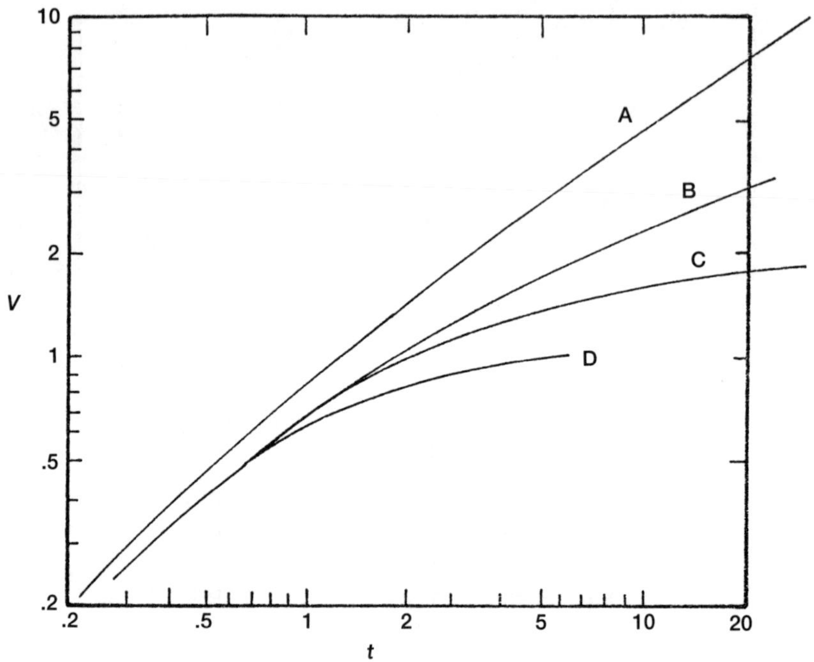

Figure 12.1. Math-model plots in constant-pressure filtration. V = cumulative volume of filtrate; t = time. Curve A reaches a slope of 0.5.

Curve	Model Name	Equation
A	Cake filtration	12.2b
B	Intermediate blocking	12.4b
C	Standard blocking	12.6b
D	Complete blocking	12.9b

versus time. It doesn't matter what units of measure we employ in this log/log plot. We then judge the rate at which our experimental curve changes slope by superimposing it over the empirical curves of reference in Figure 12.1 or in Figure 12.2.

If the experimental curve looks like Curve A, which reaches a final, straight-line slope of +0.5 in Figure 12.1, or –0.5 in Figure 12.2, we see the situation where the filter medium has not plugged, and the growing

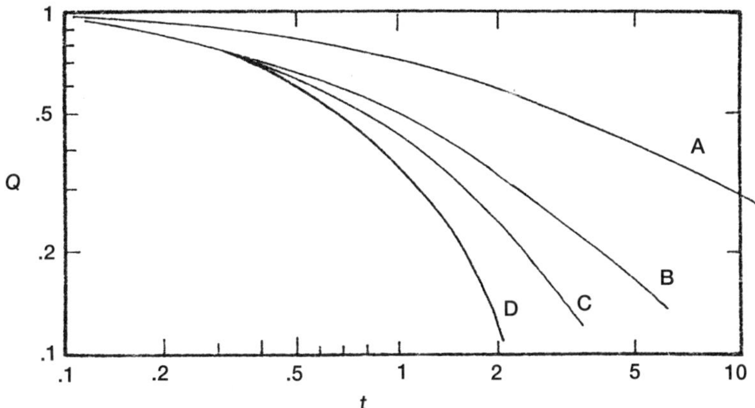

Figure 12.2. Math-model plots in constant-pressure filtration. Q = volumetric flow rate; t = time. Curve A reaches a slope of –0.5.

Curve	Model Name	Equation
A	Cake filtration	12.3b
B	Intermediate blocking	12.5b
C	Standard blocking	12.7b
D	Complete blocking	12.10b

filter cake does not compress or blind the filter medium. If the experimental curve looks somewhat like one of the other curves, we may want to try a finer filter medium and (or) the use of a filter aid.

Before explaining how to look at tests with different pressures, we show the source of the present curves of reference.

12.2 Mathematical Models in Constant-Pressure Filtration

BACKGROUND

Grace (1956), via Hermans and Bredée (1936), offers the general equation

$$\frac{d^2t}{dV^2} = K\left(\frac{dt}{dV}\right)^n$$

We consider different values of n.

12.2.1 Cake Filtration, $n = 0$

Grace writes:

$$\frac{t}{V} = \frac{K_A V}{2} + \frac{1}{Q_o} \tag{12.1}$$

where t = elapsed time,
V = volume of filtrate
K_A = rate constant, and
Q_o = initial flow rate, $(dV/dt)_o$

Equation 12.1 can also be expressed as

$$V = \left(\frac{2t}{K_A} + \frac{1}{Q_o^2 K_A^2}\right)^{0.5} - \frac{1}{Q_o K_A} \tag{12.2a}$$

that we "normalize" (see Section 12.3) in order to plot it as Curve A in Figure 12.1, as

$$V = (4t + 4)^{0.5} - 2 \tag{12.2b}$$

Grace also offers

$$\frac{Q}{Q_o} = (1 + K_A Q_o^2 t)^{-0.5} \tag{12.3a}$$

where Q = flow rate at elapsed time, t.

We normalize Equation 12.3a to

$$Q = (1 + t)^{-0.5} \tag{12.3b}$$

and plot it as Curve A in Figure 12.2.

To deduce the rate constant, K_A, consider Equation 12.1. Make a linear/linear plot of t/V versus V. The slope equals $K_A/2$.

12.2.2 Intermediate Blocking, $n = 1.0$

Graces writes

$$K_B V = \ln(1 + K_B Q_o t) \tag{12.4a}$$

that we write, for plotting in Figure 12.1, as Curve B, as

$$V = \ln(1 + t) \tag{12.4b}$$

Grace writes

$$K_B t = \frac{1}{Q} - \frac{1}{Q_o} \tag{12.5a}$$

that we write, for plotting in Figure 12.2, as Curve B, as

$$Q = (1 + t)^{-1} \tag{12.5b}$$

To deduce the rate constant, make a linear/linear plot of $1/Q$ versus t, and from the expression

$$\frac{1}{Q} = K_B t + \frac{1}{Q_o}$$

see that the slope equals K_B.

12.2.3. Standard Blocking, *n* = 3/2

Grace writes

$$\frac{t}{V} = \frac{K_C t}{2} + \frac{1}{Q_o} \tag{12.6}$$

or $$V = t\left(\frac{K_C t}{2} + \frac{1}{Q_o}\right)^{-1} \tag{12.6a}$$

that we write, in order to plot it as Curve C in Figure 12.1, as

$$V = t\left(\frac{t}{2} + 1\right)^{-1} \tag{12.6b}$$

Grace writes

$$\frac{Q}{Q_o} = \left(\frac{K_C t}{2} + 1\right)^{-2} \tag{12.7a}$$

that we write, in order to plot it as Curve C in Figure 12.2, as

$$Q = \left(\frac{t}{2} + 1\right)^{-2} \tag{12.7b}$$

To deduce the rate constant, make a plot on linear/linear coordinates of t/V versus t in order to see from Equation 12.6 that the slope equals $K_C/2$.

12.2.4 Complete Blocking, *n* = 2

Grace writes

$$V = Q_o[1 - \exp(-K_D t)] \tag{12.8a}$$

that we write as

$$V = 1 - \exp(-t) \tag{12.9b}$$

in order to plot it as Curve D in Figure 12.1.

Grace writes

$$\frac{Q}{Q_o} = \exp(-K_D t) \tag{12.10a}$$

that we write as

$$Q = \exp(-t) \tag{12.10b}$$

in order to plot it as Curve D in Figure 12.2.

To deduce the rate constant, consider

$$Q = Q_o - K_D V$$

and plot, on linear/linear coordinates, Q versus V, in order to see the slope K_D.

12.3 THE PARABOLIC CAKE-FILTRATION MODEL

BACKGROUND

This section proceeds under the assumptions, stated in Section 11.5, that the filter medium itself does not plug with solids. Or if it does the degree is negligible and the forming cake of collected solids immediately becomes the new filter medium, which may be compressible.

Under a given driving pressure, the flow rate of a fluid through a bed of collected particles is inversely proportional to the thickness of the bed. That is, as the bed grows thicker, with increased volume filtered, V, the flow rate, dV/dt, falls according to

$$dV/dt = K/V \tag{12.11}$$

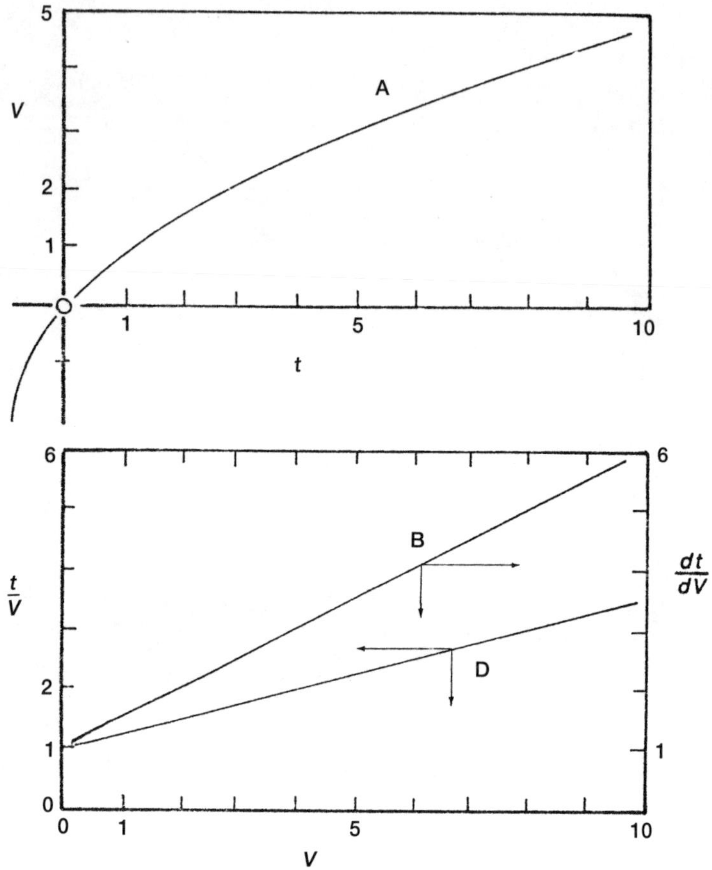

Figure 12.3. The parabola discussed in Section 12.3. Curve A is described by $(V + 2)^2 = 4(t + 1)$, or by $V = (4t + 4)^{0.5} - 2$. Line B is described by $dt/dV = V/2 + 1$. Line D is described by $t/V = V/4 + 1$.

where K is a constant. On integration, the expression becomes the parabola

$$V^2 = 2Kt + C$$

A parabola, opening to the right on a linear/linear plot of V versus t in the present concept of normalized data, passes through the origin (where $V = t = 0$) with a slope of 1.0. At the very start, V is directly proportional to

t (during the initial flow rate through the filter medium). Such a parabola has its center at $t = -1$, $V = -2$; and the expression for the curve is

$$(V + 2)^2 = 4(t + 1) \tag{12.12}$$

or, $V = (4t + 4)^{0.5} - 2$ (12.13)

Equation 12.13 is the above Equation 12.2b, and is plotted in Figure 12.3 as Curve A, but now on linear/linear coordinates. The equation can also be expressed as

$$t/V = V/4 + 1 \tag{12.14}$$

which is comparable to Equation 12.1 and, in Figure 12.3, is plotted as Line D.

Differentiating Equation 12.12 we obtain

$$dV/dt = 2/(V + 2) \tag{12.15}$$

which is our original assumption stated in Equation 12.11, but is now stated to show that at the start of filtration, where V is essentially zero, $dV/dt = 1.0$.

Equation 12.15 can also be written as

$$dt/dV = (V + 2)/2 = V/2 + 1$$

from which we make the linear/liner plot in Figure 12.3 for Line B.

The equations of Figure 12.3 are stated more thoroughly by:

$$\frac{t}{V} = \frac{V}{K} + \frac{2C}{K} \tag{12.16}$$

and

$$\frac{dt}{dV} = \frac{2V}{K} + \frac{2C}{K} \tag{12.17}$$

where $K = \dfrac{2P}{\eta c \alpha_{av}}$, and $C = \dfrac{R_m}{c \alpha_{av}}$.

t = time, $\boldsymbol{\theta}$
V = volume of filtrate per area of filter medium, $\mathbf{L^3/L^2}$
η = viscosity of filtrate, $\mathbf{F\theta/L^2}$
c = mass of solids per volume of filtrate, $\mathbf{M/L^3}$
α_{av} = average specific resistance of the cake, **L/M**
The cake is most dense and resistant next to the face of the filter medium, becoming less dense and less resistant away from the face.
P = driving pressure, $\mathbf{F/L^2}$
R_m = resistance of the filter medium, 1/**L**

In Figure 12.3, the slope of Line D is 1/K, $\mathbf{\theta/L^2}$, and the slope of Line B is 2/K, while the common intercept on the vertical axis, is 2C/K or $R_m\eta/P$, $\mathbf{\theta/L}$.

12.4 LABORATORY DEVICES FOR STUDYING CONSTANT-PRESSURE FILTRATION THE LEAF TEST

INSTRUCTIONS

Manufacturers of filtration equipmont and filter aids offer filter-testing devices, and, indeed, they offer some very elaborate setups (Dorr-Oliver, 1980). The basic device consists of a screen or *leaf* on the open end of a funnel to which a filter cloth is attached. This device is lowered into a slurry as a certain vacuum is generated on the small end of the funnel, whereupon one records the volume of filtrate with elapsed time.

The filter cloth is surrounded by a dam or ring to simulate the walls of a commercial filter, so that the filter cake will be contained.

Some devices have a knife blade for scraping off the final cake, perhaps above a precoat layer, to simulate the action of the doctor blade on a rotary-drum filter.

With some devices, the leaf is held vertically to simulate what happens in a plate-and-frame filter. Sometimes the leaf is held within a pressure vessel, so that high pressures can be studied.

Alternatively, the filter screen and cloth are fitted to the bottom of a cylinder on top of which pressure is applied.

Note that different results are obtained with different positions of the leaf because of particle sedimentation. Thus, to obtain meaningful results, the leaf must be placed in the position corresponding to that which the filter medium will be employed in a scaled-up operation.

12.5 EXAMPLE OF A LEAF TEST IN CONSTANT-PRESSURE FILTRATION

Tiller et al. (1977) provide the following example. We show their original data, then convert their units of measure to SI units for subsequent calculations via Equation 12.18. The meaning of the symbols in that equation are explained above under Equation 12.17.

Area of filter medium: 0.01 $ft^2 = 9.29 \cdot 10^{-4}$ m^2
Vacuum (pressure): 11.1 in. Hg = $3.76 \cdot 10^4$ Pa
Viscosity of filtrate: $\eta = 0.861$ cP = $0.861 \cdot 10^{-3}$ Pa • s
Mass of solids per volume of filtrate: c = 30 kg (of TiO_2)/m^3

Operating data:

time, t, sec.:	8.1	17.4	30.1	46.3	66.6
ml filtrate:	110	170	225	290	340

Data plotted in Figure 12.4:

V, m^3/m^2:	.118	.183	.242	.312	.366
t/V, s/m:	68.6	95.1	124	148	182

On inspecting the plot of t/V versus V, in Figure 12.4, we see that the slope is

$$\frac{178}{0.40} = 445 = \frac{1}{K} = \frac{\eta c \alpha_{av}}{2P} \tag{12.18}$$

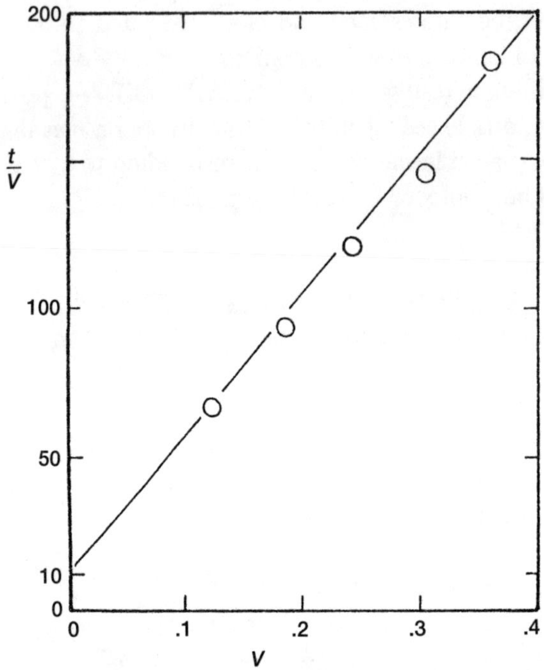

Figure 12.4. Plot of data in Section 12.5 in the manner of Line D in Figure 12.3

so that $\alpha_{av} = \dfrac{445 \cdot 2P}{\eta c} = \dfrac{445 \cdot 2 \cdot 3.76 \cdot 10^4}{0.861 \cdot 10^{-3} \cdot 30} = 1.30 \cdot 10^9$ m/kg.

Recall that α_{av} is the average specific resistance of the filter cake, *at the pressure studied.* To learn how this resistance changes with pressure, we make runs at different pressures.

Also recall, from the previous chapter, that α_{av} is a function of pressure, P, according to

$$\alpha_{av} = a(1 - n)P^n \tag{12.19}$$

where a is a constant and n is the compressibility factor (not the n of Section 12.2).

Consider another form of this equation

$$\log \alpha_{av} = n \log P + \log a + \log (1 - n) \tag{12.20}$$

Equation 12.20 tells us that on making plot of α_{av} versus P, on log/log paper, we will learn the compressibility factor, n, from the slope.

Cheape (1982), with an eye on the operation of a rotary-vacuum filter, describes a three-step leaf test in which

1. the cake is formed on dipping the leaf into the slurry,
2. the cake is washed on dipping it in water,
3. a knife, simulating a doctor blade, scraps off the washed cake.

The time allowed for each step corresponds to the rotary speed of the drum and to the distances along the circumference wherein each of the operations occur.

12.6 STERILIZATION BY FILTRATION

In studying microporous membranes designed to stop microbes, investigators address the concept of surface-versus-depth filtration. That is, if a membrane shows the symptoms of cake filtration during the separation of test microbes from a broth of such microbes, that membrane apparently stops those microbes on the surface. That is to say, the pores in the membrane are indeed small enough to stop the microbes of interest. Tanny et al. (1979) address this subject via the following tests:

Under a constant driving pressure, they fed a suspension of *Pseudomonas diminuta* to a 0.20-μm-rated membrane and separatly to a 0.45-μm-rated membrane. Both membranes, of cellulose triacetate, were apparently the same thickness and porosity, each with an area of 10 square centimeters.

With the 0.20-μm membrane, their linear/linear plot of t/V versus V (t in minutes, V in ml) showed a straight line, which, according to Equation 12.1, indicates the Cake Filtration Model.

With the 0.45-μm membrane, their linear/linear plot of t/V versus t (instead of V) showed a straight line, which, according to Equation 12.6, indicates the Standard Blocking Model.

The filtrate from the finer membrane was sterile in different runs under driving pressures ranging from 5 to 45 psi.

The coarser membrane did let some microbes through, even at the low driving pressure, which tells us that these microbes did indeed enter the depths of the membrane.

Tanny et al. conclude that the finer membrane stopped these test microbes on the surface, while the coarser membrane only stopped the microbes after they had entered the depths of the membrane. Those that were captured were captured by *absorptive sequestration.*

We now analyze their data around the 0.20-µm membrane, where they saw the symptom of cake filtration (their Figure 6). We will deduce the increased resistance of the membrane with the captured cake and then relate that resistance to the apparent density of the microbes in that cake.

In their linear/linear plot of t/V versus V, as in the present Figure 12.3, the value of t/V, at where V is essentially zero, is 0.017 min/ml. It remains at 0.017 until $V = 200$ ml, at which point it climbs as a straight line to reach 0.032 when $V = 2000$ ml. The slope of the line is

$$(0.032 - 0.017)/1800 = 8.33 \cdot 10^{-6},$$

corresponding to $K_A/2$ in Equation 12.1, that is, in the equation

$$\frac{t}{V} = \frac{K_A}{2}V + \left(\frac{t}{V}\right)_0$$

Recall from Figure 12.3 that a plot of dt/dV versus V shows twice the slope of t/V versus V, that is,

$$\frac{dt}{dV} = K_A V + \left(\frac{dt}{dV}\right)_0$$

where dt/dV corresponds to instantaneous resistance.

Thus, to reach the resistance of the membrane at the end of the run, we calculate

$$\frac{dt}{dV} = (2 \cdot 8.33 \cdot 10^{-6} \cdot 1800) + 0.017 = 0.047.$$

The resistance of the membrane, with what it gathered after 2000 ml, increased by a factor of 0.047/0.017 = 2.76.

As we reason below, not enough microbes collected on the membrane to constitute a cake. Thus, for a minute, let's not look at cake resistance (we don't know the mass of the cake), rather, let's look at pore blocking, as if the pores were blocked by the relatively few microbes on the surface.

The above increase in resistance by a factor of 2.76—or more to the point, a decrease in flow rate by a factor of 1/2.76 = 0.36—suggests only 36% of the pores remained opened and not blocked by captured microbes. That is, 64% of the pores were blocked, assuming once a microbe blocks a pore, no other microbe falls on top of it.

Now consider the numbers of microbes. These investigators fed $2 \cdot 10^7$ microbes to each square centimeter of membrane surface, as we deduce from:

$$2000 \text{ ml} \cdot (10^5 \text{ microbes/ ml}) \cdot (1/10 \text{ cm}^2) = 2 \cdot 10^7 \text{ microbes/cm}^2$$

A single, rod-shaped *Ps. diminuta*, on the surface of a membrane, will occupy an area of about $0.25 \cdot 1.0 = 0.25\ \mu\text{m}^2$. Thus, $2 \cdot 10^7$ of them, one-layer thick, will occupy an area of $0.50 \cdot 10^7\ \mu\text{m}^2$.

Assuming the membrane porosity is 0.75, so that the surface area of the pores over a membrane area of one square centimeter amounts to $0.75 \cdot 10^8 \mu\text{m}^2$, we deduce that, at most, only 6.7% of the pores can be blocked by these microbes. We deduce that from

$$(0.50 \cdot 10)^7/(0.7 \cdot 10^8) = 0.0666$$

On the other hand, if we reason that microbe count was *ten* times as high—blocking 67% of the pores—we can reason that such a high count would explain the increase in resistance. But, were the investigators really off by a factor of ten on measuring the microbe concentration in the feed stream? Probably not.

We are left to conclude that materials in the broth other than live microbes blocked the pores of the membrane, or, in keeping with our math model, formed the (apparent) permeable cake over the surface.

12.7 SOME SPECIAL TESTS OR CONCEPTS

12.7.1 The Silt-Density Index (SDI)

This index puts a label on the clarity of water via a filtration test. Water under a driving pressure of 30 psi is continually fed to a 47-mm-diameter, 0.45-µm-rated celulose-ester membrane while measuring the time for the flow of the first 500 ml, and after five minutes, the time for the flow of another 500 ml.

It's easier to explain the calculations with an example rather then trying to state a formula. Suppose the first 500 ml is collected in 0.25 minutes, and, after five minutes, the next 500 ml is collected in 0.72 minutes. In this case we calculate

$$\mathrm{SDI} = \frac{1 - \frac{0.25}{0.72}}{5} \cdot 100 = 13.0$$

The lower the index, the clearer the water (ASTM D 4189).

12.7.2 The Filterability (Filtrability) Number

K. J. Ives (in Purchas 1977, p. 292), with an eye on the filtration of water with a sand bed, suggests the expression

$$F = \frac{C_2}{C_1} \cdot \frac{h}{ut}$$

where C_2 = turbidity of the filtrate, 0.1 units in his example
C_1 = turbidity of the feed stream, 20 units
h = head loss, 2 m
u = approach velocity, 10 m/hr
t = time the filtrate remains clear, 20 hr.

In this example $F = 5.0 \cdot 10^{-5}$, a desirably low number.

12.7.3 Standard Cake-Formation Time (SCFT)

When Purchas (1977, p. 569) addresses the fact that the *thickness* of a filter cake is a function of running time and pressure (as well as the concentration of solids in the feed stream and the compressibility of those solids), he proposes a frame-of-reference *time* to form a cake *1 cm thick* under the driving pressure of interest.

12.7.4 Filterability

Melo and Tavares (1990) suggests

$$F = \frac{1}{c\alpha_{av}}$$

where (recalling the definition of the terms under the present Equation 12.17, p. 126), F is expressed in units of area. The greater this number, the more desirable the conditions.

13
Capacity of a Filter Medium in Variable-Pressure and Variable-Flow Filtration

13.1 MATHEMATICAL MODELS

INSTRUCTIONS

When a stream, feeding a filter medium, is moved by way of a centrifugal pump, the system "rides the pump curve." That is, as the filter medium loses permeability, more back pressure on the pump results in decreased flow rate.

Where a sand bed is used to clarify water, the driving force is the column of water over the bed. At the same time, flow from the bottom of the bed is controlled by a valve so that, at least at the start, the flow rate remains constant. But as the bed plugs, the valve opens more in an effort to maintain a constant flow. Eventually, however, the flow rate falls (after which the bed is backwashed and used again).

Just as we constructed the math-model plots of Figure 11.1 for constant-flow filtration and the plots of Figures 12.1 and 12.2 for constant -pressure filtration, we can construct such plots for the combination of

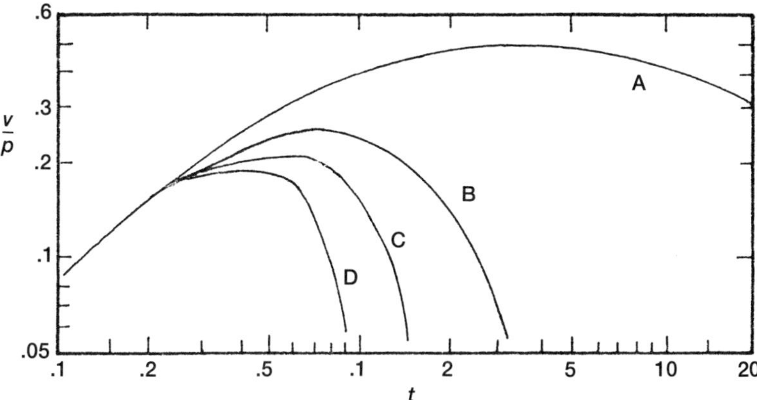

Figure 13.1. Curves of reference for variable-pressure, variable-flow filtration. V = volume of filtrate, P = driving pressure, t = time. These curves are constructed from the ratios of calculations from equations in Chapters 11 and 12.

Curve	Model Name	Equations
A	Cake filtration	12.2b/11.1b
B	Intermediate blocking	12.4b/11.3b
C	Standard blocking	12.6b/11.4b
D	Complete blocking	12.9b/11.5b

variable pressure, variable flow. We show such plots in Figures 13.1 and 13.2

Given empirical data we can plot those data on log/log paper and from that plot see a curve that appears to follow one of these math-model curves.

Tiller (1990) provides an example of filtration with a centrifugal pump, and Johnston (1992c) shows how a plot of Tiller's data follows the shapes of the *A* curves in Figure 13.1 and in Figure 13.2.

In those situations where we have done laboratory filtration tests at constant pressure, and we have done those tests at many different pressures, Tiller et al. (1977) explain how we can predict the results of employing a centrifugal pump, knowing the pressure/flow relationship of the pump.

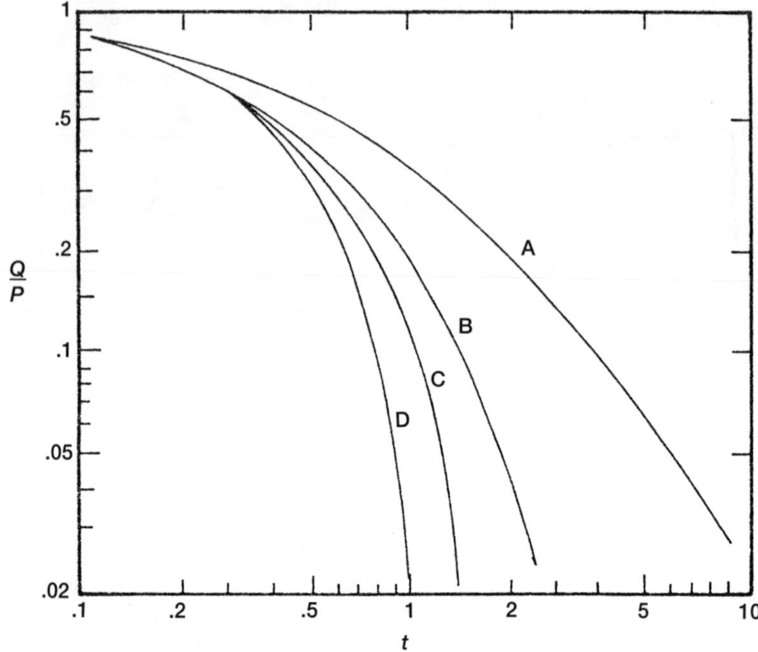

Figure 13.2. Curves of reference for variable-flow, variable-pressure filtration. Q = flow rate of filtrate, P = driving pressure, t = time. These curves are constructed from the ratios of calculations from the equations in Chapters 11 and 12.

Curve	Model Name	Equations
A	Cake Filtration Law	12.3b/11.1b
B	Intermediate Filtration	12.5b/11.3b
C	Standard Blocking	12.7b/11.4b
D	Complete Blocking	12.10b/11.5b

13.2. AN AIChE TEST PROCEDURE

Another example is provided by the American Institute of Chemical Engineers', 1967, *Equipment Testing Procedure. Batch Pressure Filters.* That procedure teaches us how to determine the resistance of the filter cake with increased driving pressure, with the complication of employing a centrifugal pump, where both the flow rate and driving pressure change with time.

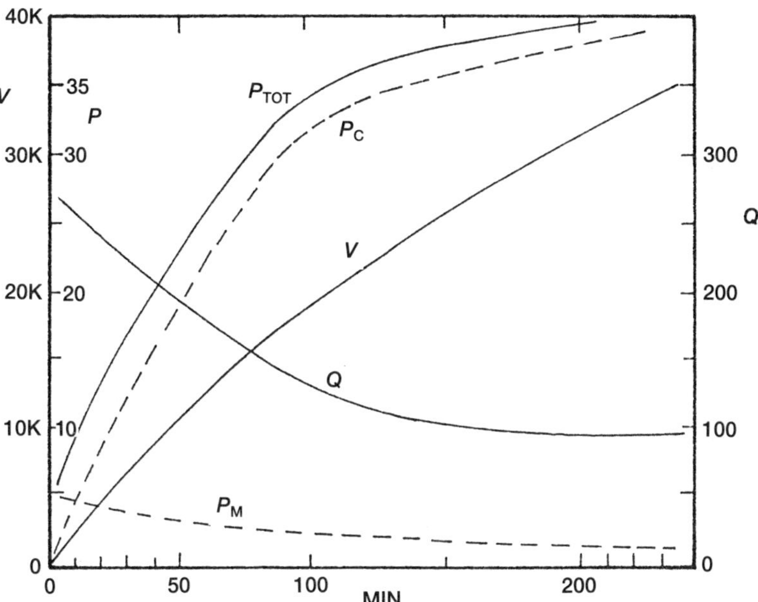

Figure 13.3. Operating data from a 500-sq-ft, plate-and-frame filter fed a talc slurry with a centrifugal pump. V = cumulative gallons of filtrate. Q = flow rate, gal/min. P_{tot} = total pressure drop across the combination filter medium and cake. P_m = pressure across the filter medium, psi, deduced from Q, and assuming the medium does not plug. P_c = pressure across the cake, deduced from $P_{tot} - P_m$.

In this example a 500-square-foot, plate-and-frame filter is fed a water slurry containing 0.003 mass fraction of talc. Thus, among the terms below, the talc concentration, c, is 0.187 lbm/ft^3.

The tabulated operating data are plotted here as the curves of Figure 13.3. The viscosity, η, of the 43 ^{0}F filtrate is 1.47 cP, equal to $3.07 \cdot 10^{-5}$

lbf • sec/ft². From Figure 13.3, we construct a short table employing the nomenclature in boldface of Chapters 11 and 12:

Time, min:	25	50	100	200
gal/min:	234	200	130	95
$\boldsymbol{u}$, 10^{-4} ft/sec:	10.4	8.92	5.84	4.24
P_C, psi across cake:	11	19.5	32	38
$\boldsymbol{P}$, lbf/ft²:	1,584	2,808	4,608	5,472
gallons:	5,800	11,000	19,500	25,000
V, ft³/ft²:	1.55	2.91	5.24	8.67

We then calculate α_{av}, the average specific resistance of the cake for each of these four different pressures in units of 10^{11} ft/lbm:

$$\alpha_{av} = \frac{P}{Vc\eta u} = 1.70 \quad 1.86 \quad 2.64 \quad 2.59$$

With more than the four calculations shown here, we clearly see that α_{av} increases with increased driving pressure. In a log/log plot of α_{av} versus P (plot not shown here), we see a slope not significantly different from 0.51, the compressibility factor of talc reported by Tiller et al. (1977).

But the authors of these AIChE data do not address the obvious questions concerning the end of the run. When the flow rate leveled off instead of dropping,

- did some of the cake fall off the filter medium?
- did a portion of the filter medium give way?

On plotting the data in the form of Figure 13.1 (plot not shown), we see that the data do begin to follow Curve A, but in the end, the test results do not fall as fast as Curve A, once they get past the hump.

Chase (1993) addresses the size of a centrifugal pump to be used.

13.3 A "DRAW-DOWN" OR RECIRCULATION SCHEME OF FILTRATION

We place this subject in this chapter simply because this kind of filtration might be performed with a centrifugal pump.

In this scheme of filtration, liquid is drawn from a container, passed through a filter, and the filtrate is continuously returned to the container

During the run both the feed stream and the filtrate are periodically examined by whatever analytical method we feel gives us what we want to know about the concentration of undesirable particles. For example, we may look at turbidity, or the concentration of *d*-diameter particles, or the mass concentration of all particles (the "gravimetric level").

Alternatively, we may only examine the feed stream to the filter, which represents the bulk of the liquid in the container when that liquid is well stirred.

While gathering the analytical data during the run, we begin making a plot where the vertical axis, a log scale, shows the concentration of whatever we measured in the feed stream.

Knowing the volume of liquid in the container and the volumetric flow rate through the filter, we construct the horizontal axis on a linear scale to show time, but marking that time scale as the numbers of "turnovers." For example, with 100 gallons of liquid in the container and with pumping 10 gal/min through the filter, we have one turnover in 10 minutes, two in 20 minutes.

If the liquid in the container is *well stirred* and filtration efficiency (however we measure it) *remains constant,* the plot will show a *straight line* with a negative slope. The slope describes the rate at which the liquid in the container becomes clearer with time. The more negative the slope, the greater the filtration efficiency.

Jim Joseph (1994) shows a plot of such theoretical conditions when the vertical scale is linear, in which case we see falling *curves* (turning out to the right).

While Hong (1985) shows log scales for the vertical axis and his plots of experimental data start out as falling straight lines, the lines then curve out to the right. Hong measures the clarity of the liquid feeding the filter by the NFPA method. That is, instead of looking at the concentration of *d-diameter* particles, he looks at the concentration of the *numbers of d-dia- meter-and-larger* particles. Hong actually follows the NFPA *multi-pass-* test procedure but without the continued addition of test dust to the feed tank. He calls his test *Beta prime.*

When the area of the filter medium is small compared to the volume of liquid to be clarified, the medium will loose permeability with time and develop a change in filtration efficiency (perhaps rising). If the loss in permeability results in significant back pressure on a centrifugal pump, then flow rate falls, in which case we must constantly monitor the flow rate to correctly lay out the horizontal axis of our plot.

14
Cross-Flow Filtration

14.1 INTRODUCTION

In cross-flow filtration the feed stream sweeps across the face of the filter medium, rather than hitting it head on. This arrangement inhibits the accumulation of particles on the medium thereby increasing its capacity.

The simplest kind of setup amounts to a column of liquid over a filter medium fitted with stirrer blades close to the upstream face of the medium. Particles that would otherwise settle on the medium are kept suspended.

Such a device is generally used to test a filter medium, usually a membrane, that is installed in the kind of device schematically described in Figure 14.1.

Figure 14.1 also represents a portion of an axial cross section of a spiral-wound, tubular cartridge (or module).

Alternatively, the membrane may be in the form of a bundle of hollow fibers. And in another arrangement, the membrane is held in a plate-and-frame kind of device; writers refer to *tangential flow* in relation to this device.

Figure 14.1 also shows that in addition to the three *streams,* we have four different *pressures* to consider. In correctly operated units, the four different pressures fall according to the increased identifying numbers shown. If P_2 is not greater than P_4 , then full use is not being made of the surface of the filter medium.

14.2 VOCABULARY

To address cross-flow filtration, we must add to—and, in some cases, change—our vocabulary. Stream Q_F in Figure 14.1 is still called the feed stream. Stream Q_C is, logically, called the *concentrate*, and sometimes the *retentate*. Stream Q_P is called the *permeate* (instead of the filtrate).

The *velocity* of the liquid leaving the face of the membrane (having passed through the membrane as permeate) is called *flux*, with the symbol J. (Does *flux* sound more scientific, even though an earlier meaning is dysentery?). Yet, because the word is not as specific as velocity, we continue to see and hear the use of "flux rate" (akin to the landlubberish "knots per hour").

Further, we also see the redundant "filtrate flux."

Filtration efficiency is referred to as *rejection,* or the *rejection rate,* or the *sieving ratio*.

In the United States, the units of J are usually expressed as gal/day per square foot—the kind of units meant for the person who must deduce the area of membrane needed for the desired size of the permeate stream. Yet, *not* accompanying such reports are sizes of the other streams, the feed and the concentrate.

Conversion refers to the flow ratio: permeate/feed.

14.3 FLOW RATIOS OF THE THREE STREAMS

What happens to the concentrate stream? In some cases it is recirculated back to the source of the feed stream.

For example; when a fermenter is growing microbes, a cross-flow-filter module is employed in two ways:

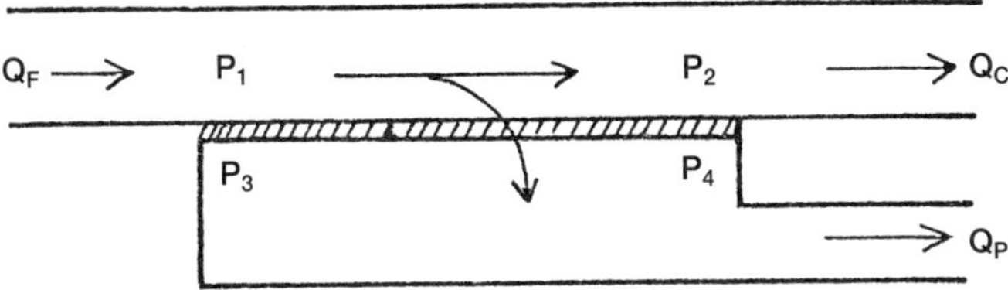

Figure 14.1. Schematic diagram of a setup in cross-flow filtration. Q = volumetric flow rates: Q_F = feed stream; Q_C = concentrate; Q_P = permeate, for an area of filter medium indicated by the cross-hatched line. Ideally, pressures, *P*, decline in the order of increasing numbers shown. That is, when P_2 is not greater than P_4 , then full use is not being of the surface of the medium.

1. During the growth of the microbes, a stream from the fermenter is fed to the module as soluble waste products pass through the membrane (as permeate), and the microbes return to the fermenter in the concentrate stream. Meanwhile, fresh nutrients are added to the fermenter.
2. The final microbe concentration in the fermenter is increased by recirculating the liquid through the cross-flow module, during which time the water, with its soluble material, is separated via the permeate.

Sometimes the concentrate is not recirculated to the source of the feed stream. When a module is used to reduce the salt content of water, the concentrate is discarded, while the permeate becomes the product of the operation.

Before we address the ratio of permeate flow to feed flow, or to concentrate flow, we will first describe the different kinds of membranes employed.

14.4 CLASSIFICATIONS OF MEMBRANES

When microbes or microbe-sized particles (diameters: 0.02 to 10 μm) are to be separated, the process is called *microfiltration.*

When large-molecular-weight and soluble materials, such as sugars and proteins (diameters: 0.001 to 0.02 μm, or 10 to 200 Angstroms), are to be separated the process is called *ultrafiltration.*

And when soluble salts (diameters: 1 to 10 Angstroms) are to be separated the process is called *reverse osmosis* (Porter 1979).

Obviously, each of the above processes calls for a membrane with a specific flow-average pore size. Yet, in the case of reverse osmosis—and, to some extent, in ultrafiltration—pore size is not as important as the simple nature of the membrane, meaning the material of construction.

Further, the finer the membrane (the smaller the pores), the thinner the membrane, which enables a reasonable flux to be obtained with a reasonable driving pressure. The very fine and thin membranes are manufactured as a "skin" on top of a coarse, support membrane. Alternatively, some "membranes" are formed in place, within the module, through the process of laying down a precoat of fine material on top of a coarser medium.

14.5 FLUX DECAY

Consider a set of operating conditions, to wit:

- the pressure drop across (through) the membrane,
- the velocity of the feed stream across the feed-stream face of the membrane, and
- the concentration of materials in the feed stream relative to that in the permeate.

Given a certain membrane, and the materials to be separated from the fluid, the membrane will lose permeability with time, and the flux will decay. In such a case, the membrane can often be backwashed and used again. But back washing may not restore *all* of the original permeability. In commercial operations, cross-flow modules are routinely backwashed, and repeatedly used again— but only for a finite number of cycles.

The required frequency of the backwash operation depends on many factors. And, indeed, the frequency may not be all that frequent if the investigator can find the proper conditions and the proper membrane.

14.6 TEST PROCEDURES

Given the geometry of the module, the membrane in place, and the fluid to be filtered, only two variables remain:

- the pressure drop across (through) the membrane, and
- the velocity of the fluid over the feed-stream face of the membrane.

Changing one without the other is done via changes the feed-stream pressure, and with changes how severe the two exit lines are separately pinched down.

Yet pressure gauges are *not* located within the module, as implied by Figure 14.1. Instead they are located in the streams feeding and emitting from the module, that is, relatively far from the membrane. See, for example, the diagram of a reverse-osmosis test procedure in ASTM D 4516.

Thus, since we really don't know the drop in pressure in the lines carrying those streams, statements about the pressure drop through the membrane, and the pressure drop across the face of the membrane, along the feed-stream channel, are made partly tongue in cheek. Indeed, this author has yet to see another writer assure his or her readers that, in the diagram in Figure 14.1, the four pressures identified do indeed descend in the order they are identified with the increasing identification numbers.

What investigators really measure are the differences in pressures among the three different streams. Indeed, all we can do is hunt for the pressure of each stream, measured close to the module, that will yield the best results.

In the literature, cross-membrane flow (along the feed channel) is referred to as shear rate, γ (1/sec), which is deduced from velocity, u, via either of the following two kinds of calculations:

- for hollow fibers, of diameter, d, $\gamma = 8u/d$
- for rectangular slits, of height, h, $\gamma = 6u/h$.

And in the literature we see plots, on log/log paper, of flux versus shear rate (they rise together). Yet the authors fail to state whether the plots are made with a constant pressure drop through the membrane, or not.

14.7 THE BOUNDARY LAYER; MORE ON FLUX DECAY

As stated at the top of this chapter, the high velocity of the feed stream over the membrane inhibits the accumulation of solids. Nonetheless, a boundary layer does form. Ideally, we look for a layer that is thin, remains constant, and doesn't build up with time.

Further, we look *only* for a layer; that is, the pores in the membrane do not plug.

The composition of the layer obviously depends on what's in the fluid. Whether or not the pores in the membrane become plugged depends on the pore sizes relative to the sizes of the solids in the feed stream, and on the filtration efficiency (rejection rate).

Sometimes the boundary layer is a simple cake, as addressed in chapters 11–13. Other times it is so dense that the velocity (flux) of water (or solution) through it is controlled by the liquid *diffusion rate*. That is, no amount of extra pressure on the membrane will increase the flux.

As with a "dead end" filter cartridge, a cross-flow module may be used in a batch operation, in which case "life" or capacity is not as important as it is in continuous use.

14.8 EXAMPLES

14.8.1 Reverse Osmosis

The pressure drop through the membrane is obviously important in reverse osmosis, since it is only this pressure drop that provides the salt-rejection rate (to overcome back, osmotic pressure). That is to say, in converting sea water to drinking water, high pressures are required. Yet relatively low pressures can be employed to convert brackish water to drinking water. Indeed, in some communities where the sodium level is high, home-use, reverse-osmosis units employ water-line pressure to do the job, but at the expense of low *conversions.*

For example, a home unit is built to operate with, say, a 40-psi feed stream, to produce 5 gallons per day of drinking water while discharging perhaps 30 gallons per day to the sewer. The *conversion* is 5/30 = 0.17 (17%) (ASTM D 4195, D 4516).

14.8.2 Harvesting Microbes

A veterinary vaccine is produced by growing *Streptococus pyogenes* in a 286-liter soup. After the organisms are grown to the maximum population, the mixture must be concentrated by a factor of 13 to be in the proper dosage form.

To increase this concentration, a hollow-fiber module is used, containing 1 square foot of membrane surface (0.093 m^2) with a pore-size *rating* of 0.2 μm. Using a positive-displacement, lobe-type pump, the mixture is fed to the module at a pressure of 15 psi (1.0 bar) with outlet lines unrestricted (zero gauge pressure).

The initial permeate flow of 3.5 liters/min drops to 1.0 liters/min in about an hour, where it essentially remains for the next 3.5 hours (with a slight decay). Meanwhile the concentrate stream, returning to the fermenter, also flows at an average rate of 1.0 liters/min.

In 4.5 hours, with the permeate volume reaching 264 liters, the original microbe concentration in the fermenter has been increased by a factor of 286/(286 – 264) = 13 (Norquist 1987).

14.8.3 Plasmapheresis

In plasmapheresis, blood, under an initial driving pressure of nearly 1 mm Hg, is fed to a 0.5-square-meter, hollow fiber module (whose pore-size rating is not reported but which is designed for this procedure) at the rate of 100 liters/minute. The resulting flow of plasma as the *permeate* amounts to 5 ml/min, with the remaining 95 ml/min of *concentrate* returning to the patient.

Over a period of four hours the feed pressure is increased, stepwise, to 35 mm Hg, while the feed rate is maintained at 100 ml/min. Such stepwise action increases the flow of plasma up to 35 ml/min. (Werynaski et al. 1981; Malchesky et al. 1989).

14.8.4 Cross-Flow Electrofiltration

In one arrangement, a module, consisting of a single, hollow tube with porous walls, functions as an electrode through which clarified water, or oil, passes. The other electrode is the wall of the module.

As a direct-current potential across the electrodes is applied, particles in the feed stream, flowing past the porous tube, are repelled. Hence, the face of the porous tube does not experience the buildup of a boundary layer.

Either water with suspended solids or oil with suspended solids can be treated, with more applied voltage for the oil than for the water. But when the oil contains too many water droplets, the electrodes tend to short out. The applied polarity across the electrodes depends on the zeta potential of the suspended solids (Verdegan et al. 1985; Verdegan 1986).

15
Oil-Water Separation

15.1 INTRODUCTION

This unit operation is commercially important in separating water from fuel and lube oil, and in separating oil from a ship's bilge water before that water is pumped over the side (Verdegan and Jaisinghani 1980).

The relatively simple operation of detaching two separate *layers* from a mixture (oil floats on water) is performed by a unit appropriately called a *separator*, or *decanter*, the design of which is based on the simple difference in densities between the two liquids, and the flow rate of the feed stream.

But when the liquid of the *discontinuous* phase is dispersed in the other liquid of the *continuous* phase, as a stable emulsion, the small droplets of the discontinuous phase must be made to join together into larger and more easily separated drops, via the passing of the emulsion through a *coalescer* (Jaisinghani et al. 1977; Verdegan 1987).

The coalescer, usually a mat of fibers, need not be very thick. Thin fibers are preferred over fat ones; and, they need not be packed too close together—after all, one wants a low hydraulic pressure drop across the mat.

Alternatively, the emulsion can be passed between corrugated plates. That is, the stream, moving horizontally, is forced to travel an up-and-down, sine-curve route (Jaisinghani and Sprenger 1979).

Troublesome to either operation is the presence of suspended solids. Those solids can quickly plug a fibrous-mat coalescer. And in the case of the decanter, solids may be incorporated into the oil layer, making that layer so dense that it no longer floats on water. Difficulties also arise when the oil is very viscous.

15.2 TEST METHODS

Apparently, no *standard* test methods exist for evaluating a coalescer because of the many variables involved, such as those addressed by the following questions: (Jaisinghani 1977; Verdegan and Jaisinghani 1980; Jaisinghani and Verdegan 1982)

- Should the water phase contain salt (NaCl)? If so, how much?
- Which viscosity of oil do we employ in putting together a test emulsion?
- Should the oil contain one of the many usual additives present in oils (Jaisinghani 1977)?
- What surfactants, if any, do we employ? That is, what sort of *interfacial tensions* and *interfacial-shear viscosities* do we build?
- How do we address electrokinetic effects?
- Do we include test particles? If so, what size, and what concentration?

Bibliography

Some of the papers cited here, kindly sent to us as a result of our plea for test methods, are unpublished manuscripts, or even a collection of notes. The copies of some published papers came to us without a complete citation. And some papers published as notes of a company selling services came to us without dates. That is to say, the reader will see some incompleteness in our citations, for which we apologize.

Abbott, Edwin A. 1963. *Flatland. A Romance of Many Dimensions,* 5th ed. New York: Barnes & Noble.

AIChE. 1967. *Equipment Testing Procedure Batch Pressure Filters.* New York

Alderete, J. 1991. "A Practical Approach to Filtration Ratings & Effective Filter Selection," Cuno, Inc. memo DOR 1606, Meriden, Connecticut

ASTM D 3862. *Retention Characterisrics of 0.2-µm Membrane Filters Used on Routine Filtration Procedures for the Evaluation of Microbiological Water Quality.* This procedure, and the other ASTM procedures to follow, appear in *1991 Annual Book of ASTM Standards.* Philadelphia.

_____. D 3863 *Retention Characterisrics of 0.40 to 0.45-µm Membrane Filters Used on Routine Filtration Procedures for the Evaluation of Microbiological Water Quality.*

_____. D 4189. *Silt Density Index of Water.*
_____. D4194. *Operating Characteristics for Reverse Osmosis Devices*
_____. D4195. *Water Analysis for Reverse Osmosis Application.*
_____. D4516. *Practice for Standardizing Reverse Osmosis Performance Data.*
_____. E128. *Maximum Pore Diameter/Permeability of Rigid Filters.*
_____. F316. *Pore Size Characteristics of Membrane Filters by Bubble Point and Mean Flow Pore Test.*
_____. F660. *Particle Size Comparisons in Use of Alternative Particle Counters.*
_____. F778. *Gas Flow Resistance Testing of Filtration Media.*
_____. F795. *Performance Testing of Filter Medium Using a Single-Pass, Constant-Rate Liquid Test.*
_____. F796. *Performance Testing of Filter Medium Using a Single-Pass, Constant-Pressure, Liquid Test.*
_____. F797. *Performance Testing of Filter Medium Using a Multi-Pass, Constant-Rate Liquid Test.*
_____. F838. *Bacterial Retention of Membrane Filters Utilized for Liquid Filtration.*
_____. F902. *Average Circular-Capillary Equivalent Pore Diameter in Filter Media.*
_____. F1067. *Performance of Petroleum Product Filter Cartridge, Using a Single-Pass, Constant-Rate Procedure.*
_____. F1215. *Initial Particle Size Efficiency of Flatsheet Medium in an Airflow, Using Latex Spheres.*
_____. F1170. *Performance of Filter Medium Using Water and Siliceous Particles.*
_____. F1260. *Liquid Drop Size Characteristics in Sprays Using Optical Non-Imaging Light-Scattering Instruments*
_____. 1986. STP [Special Technical Publication] 975, *Fluid Filtration. Vol. I, Gas; Vol. II, Liquid.*

Badenhop, C. Thomas. 1983. "The Determination of the Pore Distribution and the Consideration of Methods Leading to the Prediction of Retention Characteristics of Membrane Filters," Dr. Ing. Disertation, University of Dortmund, Germany.

Bader, H. 1970. "Hyperbolic Distribution of Particle Sizes." *J. Geophysical Research* 75: 2822–2830.

Baumann, E. Robert. a. "Filter Dynamics," MS outline.

_____ . b. "Testing—System Characteristics," MS outline.

_____ . c. "Solid-Liquid Separation," MS outline.

_____ . d. "Testing—Current State of the Art," MS outline.

Bear, Jacob. 1972. *Dynamics of Fluids in Porous Media.* New York: American Elsevier.

Bently, J. M., and P. J. Floyd. 1992. "Interpreting the Rating of Cartridge Filters," *Filtration & Separation,* July/Aug.: 333–335.

Blunt, William G. 1992. "Procedure for Evaluating Permeability of Disposable Filter Media Using Bomb Filter Apparatus." Harborlite Corp. Bulletin 292A.

Brandt, Robert H. 1972. "The Analysis of Particulate in 'Filtered' Coolant." *J. American Soc. of Lubrication Engineers*, July: 254–257.

Buffington & Associates, "Baghouse Systems and Components." MS.

Campbell, J. S., and M. Iwanaga. 1981. "Beta Rating Variation with Different Test Contaminants." *The BFPR Basic Fluid Power Research Program Journal* 14: 87–93.

Carman, P.C. 1937. "Fluid Flow through Granular Beds." *Transactions Institute of Chemical Engineers London* 15: 150–166.

_____ .1956. *Flow of Gases through Porous Media.* London: Butterworths

Chambers, Catherine D, and Thomas A. Janszen. 1990. "Point-of-Entry Drinking Water Treatment Systems for Superfund Applications" EPA Project Summary, Cincinnati.

Chase, George G. 1993. "Accounting for Pipe Loss when Compairing Centrifugal Pumps for Your Filtration," *Fluid/Particle Separation J.* 6: 84–89.

Cheap, Dudley W. 1982. "Leaf Tests Can Establish Optimum Rotary-Vacuum-Filter Operation," *Chemical. Engineering,* June 14, 141–148.

Cheremisinoff, Nicholas B., and David Azbel. 1983. *Liquid Filtration,* Woburn, Mass.: Ann Arbor Science Publishers.

Chiang, Shiao-Hung. 1992. "Cake Filtration and Dewatering." MS outline.

Chrys, Philip Z. 1991. "Improving Quality and Productivity Through Filter Media Selection" SME Technical Paper S91-267: Dearborn, Mich.

_____ . 1993. "Determination of the Particulate Level and Emulsion Stability of Water Soluble Cutting Fluids." MS, Lydall, Inc.

Clarke. Stephen J. "Filtration in the Sugar Industry" MS, Audubon Sugar Institute.

Cole, Fred. 1966. "Particle Count Rationalization." Bendix Filter Div.

Conner, W. C., A. M. Lane, and A. J. Hoffman. 1984. "Measurements of the Morphology of High Surface Area Solids: Histeresis in Mercury Porosimetry." *J. Colliod and Interfacial Science* 100: 186–193.

Coyne, K., W. C. Conner, and K. Rucinski. 1986. "Filter Morphology and Performance: Porosity and Microscopy of Oil GFilter Media Compared with Filtration." *The Chemical Engineer* 32: 53–62.

Davies, C. N. 1973. *Air Filtration*. New York: Academic Press.

De Bruyne, R. "The Application of 'Non media' Standards in the Design and Testing of Filter Media." From a publication of the Belgian Filtration Society.

______. 1991. "Definition of Bekipor ST Characteristics and Test Methods." Bulletin of Bekaert Fibre Technologies.

Dickenson, Christopher. 1992. *Filters and Filtration Handbook*. Oxford, England: Elsevier Science Publishing Co.

Dorr-Oliver Co. 1980. *Selecting and Sizing Dorr-Oliver Filters. Filtration Leaf Test Procedures.*

Dullien, F. A. L. 1979. *Porous Media Fluid Transport and Pore Structure*. New York: Academic Press, pp. 159–161.

Edyvean, R. G. J., and C. J. Williams. "Testing Cartridge Filters/Interpreting Results— Pitfalls and Problems." MS, University of Leeds.

Eggerstedt, Paul. 1991. "Solid-Liquid Separation Unique bench filter." MS, Industrial Filter & Pump Mfg. Co.

Ehlert, Joerg, and Garen Evans. "The ANDRITZ Ruthner, Inc. Hyperbaric Filter," MS.

Ergun, Sabin. 1952. "Fluid Flow Through Packing Columns." *Chemical Engineering Progress* 48: 89–94.

Fallon, Stephen L., Barry M. Verdegan, and Robert F. Ring. 1990. "Reducing Corrosion and Wear in Digester Gas-Fired Engines." 63rd Conference of the Water Pollution Control Federation, Washington, D. C.

Fitch, R. C. 1970. "Supplemental Report 70-1 of the Basic Fluid Power Research Program Annual Report." Oklahoma State University, Stillwater.

Francis, Tom. 1992. "Arizona Test Dust Update—October 1992." Powder Technology Inc., Burnsville, Minn.

Grace, H. P. 1956. "Structure and Performance of Filter Media," *AIChE Journal* 2: 307–336.

Grant, Donald C., 1988, "Sieving Capture of Particles by Microporous Membranes Filtration Media," Master's thesis, University of Minnesota, Minneapolis.

Grant, Donald, and J. G. Zahka. 1990. "Sieving Capture of Particles by Microporous Membrane Filters from Clean Liquids." *Swiss Contamination Control* 3: 160–164.

Green, L, and P. Duwez. 1951. "Fluid Flow through Porous Metals." *J. Applied Mechanics* 18: 39–45.

Haring, R. E., and R. A. Greenkorn. 1970. "A Statistical Model of a Porous Medium with Nonuniform Pores." *AIChE Journal* 16: 477–483.

Hermans, P. H., and H. L. Bredée. 1936. "Principles of the Mathematical Treatment of Constant-Pressure Filtration." *J. Soc. Chemical Industry* 55: T1–4.

Heywood, H. 1969. "Particle Characterization Conference," Loughborough University of Technology. Notes published by Coulter Electronics, Inc., Hialeah, Florida.

Hong, I. T. 1985. "The Beta Prime—a New Advanced Filrtration Theory." *Filtration & Separation,* July/Aug.., 235–238.

Jacobs, Louis J. 1984. "Filtration." In *Perry's Chemical Engineers Hand book,* 6th ed. New York: McGraw-Hill.

Jaisinghani, Rajan A. 1977. "Effect of Commercial Fuel Oil Additives on Coalescer Performance." *Filtration & Separation,* July/Aug.

Jaisinghani, Rajan A.; C. Wilkie; R. Roberts; G. Sprenger, and G. Deeley. 1977. "Design Aspects of Fibrous Bed Coalescers." Conference on Theory, Practice,and Process Principles for Physical Separations. Asilomar Conference Grounds, California.

Jaisinghani, Rajan A., and G. S. Sprenger. 1979. "A Study of Oil/Water Separation in Corrugated Plate Separators." *J. of Engineering for Industry* 101: 441–448.

Jaisinghani, Rajan A. , and Barry M. Verdegan. 1982. "Electrokinetics in Hydraulic Oil Filtration—The Role of Anti-Static Additives" World Filtration Congress III.

Jaroszczyk, Tadeusz. 1987a. "Vortex Dust Feeder for Industrial Re-

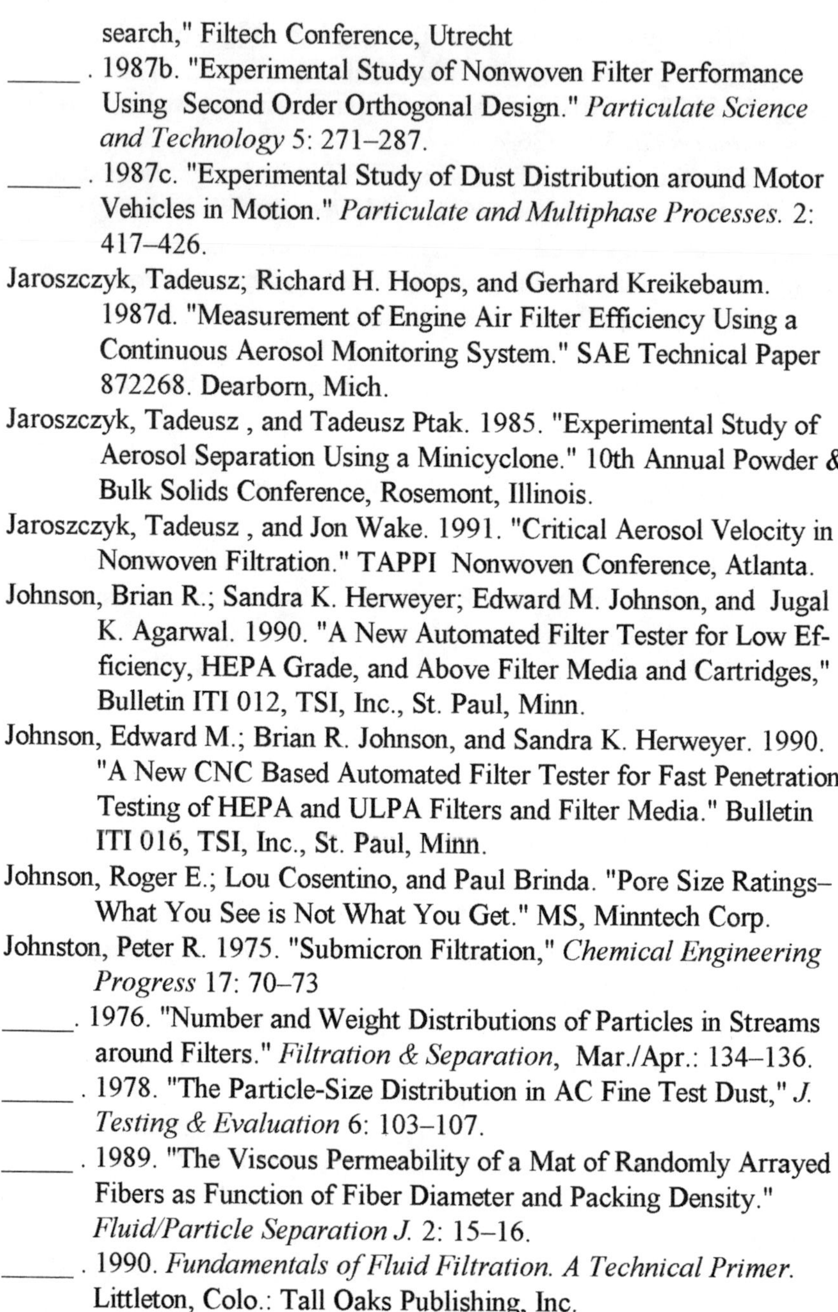

search," Filtech Conference, Utrecht

_____ . 1987b. "Experimental Study of Nonwoven Filter Performance Using Second Order Orthogonal Design." *Particulate Science and Technology* 5: 271–287.

_____ . 1987c. "Experimental Study of Dust Distribution around Motor Vehicles in Motion." *Particulate and Multiphase Processes.* 2: 417–426.

Jaroszczyk, Tadeusz; Richard H. Hoops, and Gerhard Kreikebaum. 1987d. "Measurement of Engine Air Filter Efficiency Using a Continuous Aerosol Monitoring System." SAE Technical Paper 872268. Dearborn, Mich.

Jaroszczyk, Tadeusz , and Tadeusz Ptak. 1985. "Experimental Study of Aerosol Separation Using a Minicyclone." 10th Annual Powder & Bulk Solids Conference, Rosemont, Illinois.

Jaroszczyk, Tadeusz , and Jon Wake. 1991. "Critical Aerosol Velocity in Nonwoven Filtration." TAPPI Nonwoven Conference, Atlanta.

Johnson, Brian R.; Sandra K. Herweyer; Edward M. Johnson, and Jugal K. Agarwal. 1990. "A New Automated Filter Tester for Low Efficiency, HEPA Grade, and Above Filter Media and Cartridges," Bulletin ITI 012, TSI, Inc., St. Paul, Minn.

Johnson, Edward M.; Brian R. Johnson, and Sandra K. Herweyer. 1990. "A New CNC Based Automated Filter Tester for Fast Penetration Testing of HEPA and ULPA Filters and Filter Media." Bulletin ITI 016, TSI, Inc., St. Paul, Minn.

Johnson, Roger E.; Lou Cosentino, and Paul Brinda. "Pore Size Ratings– What You See is Not What You Get." MS, Minntech Corp.

Johnston, Peter R. 1975. "Submicron Filtration," *Chemical Engineering Progress* 17: 70–73

_____. 1976. "Number and Weight Distributions of Particles in Streams around Filters." *Filtration & Separation*, Mar./Apr.: 134–136.

_____ . 1978. "The Particle-Size Distribution in AC Fine Test Dust," *J. Testing & Evaluation* 6: 103–107.

_____ . 1989. "The Viscous Permeability of a Mat of Randomly Arrayed Fibers as Function of Fiber Diameter and Packing Density." *Fluid/Particle Separation J.* 2: 15–16.

_____ . 1990. *Fundamentals of Fluid Filtration. A Technical Primer.* Littleton, Colo.: Tall Oaks Publishing, Inc.

_____ . 1991. "Pore Sizes in Filter Media," MS, American Filtration Society Conference on the pore, Hershey, Pa.

_____ . 1992a. "Pore Size and Distribution of Sizes." In *Encyclopedia of Chemical Processing & Design,* Vol. 21. John McKetta (Ed.), New York: Marcel Dekker.

_____ . 1992b. "Fluid Flow through Filter Media, A Tutorial." *Pharm. Tech.* 16: 136–146.

_____ . 1992c. *Fluid Sterilization by Filtration.* Buffalo Grove, Ill.: Interpharm Press

Johnston, Peter R., and James E. Schmitz. 1974. "A New and Recommended Way to View the Test Performance of Cartridge Filters." *Filtration & Separation,* Nov./Dec., 581–585.

Johnston, Peter R., and Roy Swanson. 1982. "A Correlation Between the Results of Different Instruments used to Determine the Particle-Size Distribution in AC Fine Test Dust." *Powder Technology* 32: 119–124.

Jones, Stuart L. "Filterability Testing as a Process Control Tool for Filter Media Manufacturing." MS, Eastman Kodak Co.

Joseph, James J. 1994. "Pilot Testing Liquid Clarification Equipemt by Progressive Dilution." *Filtration News,* Mar./Apr., 44–45.

Kaplan, Louis A., and Thomas L. Bott. 1990. "Nutrients for Bacterial Growth in Drinking Water: Bioassay Evaluation." EPA Project Summary, Cincinnati, Ohio.

H. Lindstrom, and H. Tore. "Validation of Integrity Test Values for Cleanable Porous Stainless Steel Polymer Filters," MS, Pall Corp.

Macdonald, I. F.; M. S. El-Sayed; K. Mow, and F. A. L. Dullen. 1979. "Flow Through Porous Media—The Ergun Equation Revisited," *Industrial Engineering Fundamentals* 18: 199–208.

Malchesky, Paul S., 1984, "Membrane Plasma Separation: Critical Isues," In *Therapeutic Apheresis: A Critical Look.* Cleveland: ISAO Press

Malchesky, Paul S.; T. Horiuchi; J. J. Lewandowski, and Y. Nos. 1989. "Membrane Plasma Separation and the On-Line Treatment of Plasma by Membranes." *J. Membrane Science* 44: 55–88.

Malchesky, Paul S.; Jeffrey Starre; Steven J. Subichin; Jan J. Lewandowski, and Yukihiko Nos. 1985. "Modeling of Membrane Plasma Separaion." In *Progress in Artifical Organs—1985,* Cleveland: ISAO Press.

Malchesky, Paul S.; J. Wojcicki, and Y. Nos. 1983. "Membrane Effects in Plasma Separation." In *Progress in Artifical Organs—1983.* Cleveland: ISAO Press.

Mayer, Ernest. 1992a. "Filter Aid Testing Methodology." Prepared for American Filtration Society meeting.

_____ . 1992b. "Actual Challenge Testing vs. Micron Rating and Coulter Porometer Values," Prepared for American Filtration Society meeting.

_____ . 1993. "Actual Challange Testing vs. Micron Rating and Coulter Porometer Values." *Fluid/Particle Separation J.* 6: 58–63.

McBroom, Kendall. 1993. Conversation with Peter Johnston

Melling, John. 1992. "Selection of Cartridge Filters for Clarification of Liquids." Prepared for a meeting of The Filtration Society.

Melo, Freitas M. A., and F. A. Tavares. 1990. "Filterability of Raw Sugar: Influence of Filter Aides." *Fluid/Particle Separation J.* 3: 96-100.

Meltzer, Theodore H. 1986. *Filtration in the Pharmaceutical Industry.* New York, Marcel Dekker.

Mertens, B., and A Huyghebaert. 1989. "Particle Size Analysis of Emulsions. Evaluation of Methods," *Filtration & Separation,* Sep./Oct., 352–355.

Meyer, Brad A., and Darrell W. Smith. 1985. "Flow through Porous Media: Comparison of Consolidated and Unconsolidated Materials," *I&EC Fundamentals* 24: 360–368.

Miller, Bernard, and Ilya Tyomkin. 1986. "An Extended Range Liquid Extrusion Method for Determining Pore Size Distributions." *Textile Research J.* 56: 35–40.

Mugg, Daniel J., and Kevin P. Nicolaizo. 1989. "Filtration Laboratory Test Procedure Gram Life Efficiency." MS, Cuno, Inc. , Meridan, Conn.

Nash, P. M.; P. S. Malchesky, and P. Chandhoke. 1977. "An Efficient, Compact and Simple-To-Use Blood Gas Exchanger for Long-Term Use," *Transactions Amerrican Society of Internal Organs.* XXIII: 479–489.

NFPA. 1990. ANSI/NFPA T3.10.8.8 RI. *Hydraulic Fluid Power Filters. Multi-Pass Method for Evaluating Filtration Performance*. Milwaukee.

Norquist, Roger. 1987. "Solving Liquid/Solids Separation Problems with Hollow Fiber, Cross-Flow Microfiltration." In *Pharmaceutical Filtrations*. Dearborn, Mich.: Society of Manufacturing Engineers.

Omokawa, Susumu; P. S. Malchesky; James B. Goldcamo; Susan R. Savon, and Y. Nos. 1991. "Immunomodulating Effects of Serum-Material Interactions" *J. Biomedical Materials Research* 25: 621–636.

Piekaar, H. W., and L. A. Clarenburg. 1967. "Aerosol Filters—Pore Size Distribution in Fibrous Filters." *Chemical Engineering Science* 22: 1399–1408.

Pierce, M.; R. Guimond, and N. Lifshutz. "A General Correlation of DOP Penetration with Face Velocity as Function of Particle Size Using the FTS-200." Bulletin ITI 014, TSDI, Inc. St Paul, Minn.

Porter, M. C. 1979 "Membrane Filtration." In *Separation Technology for Chemical Engineers*. P. A. Schweitzer (Ed.), New York: McGraw-Hill.

Purchas, Derek B. 1977. *Solid/Liquid Separation Equipment Scale-up.* Croydon, England: Uplands Press.

Pyle, Bobby E., and Joseph D. McCain. 1987. "Critical Review of Open Source Particulate Emission Measurements: Field Comparison," EPA Project Summary, Research Triangle Park, North Carolina.

Remiarz, Richard J.; Brian R. Johnson, and Jugal K. Agarwal. "Automated Systems for Filter Efficiency Measurements," Bulletin ITI 002, TSI, Inc., St. Paul, Minn.

Rosenstein, N. D.; A. Dybbs, and R. V. Edwards. 1980. *Non Linear Laminar Flow in a Porous Medium.* Publication FTAS/TR-80-148, Dept. of Mechanical and Aerospace Engineering, Case Western Reserve University, Cleveland, Ohio.

Reti, Adrian R.. 1977. "An Assessment of Test Criteria in Evaluating the Performance and Integrity of Sterilizing Filters." *Bull. Parenteral Drug Assoc.* 31: 187–194.

Rushton, A., and P. V. R. Griffiths. 1977. In Chapter 3 of *Filtration Principles and Practice, Part I.* Clyde Orr (Ed.), New York: Marcel Dekker.

SAE. 1988. J1858. Full Flow Lubricating Oil Filters Multipass Method for Evaluating Filtration Performance. Warrendale, Pa.

Scheuermann, E. A. 1991, "Membrane Filter, Specific Liquid Flow Rate for Flat Filter Testing." DIN 58355.

Scheidegger, Adrian. 1963. *The Physics of Flow Through Porous Media.* Toronto: University of Toronto Press.

Schwandt, Brian W., and Barry Verdegan. 1990. "Influence of Lube Oil Filter Performance on Engine Wear in City Buses." Society of Automotive Engineers Technical Paper 902238. Warrendale, Pa.

Smith, Donald R. "Acceptance Testing of Filter Media in Liquid Filtration," MS, Alcoa Tech. Center.

Simpson, K. L., and S. G. Iverson. 1989. "A Comparison of Two NaCl Test Systems Used for Measuring Filter Efficiency." Bulletin ITI 017, TSI, Inc., St. Paul, Minn.

Stinavage, Paul. 1991. "Filter Qualification and Validation," Pall Corp.

Stinson, Jeffrey A.; Matthew N. Meyers; Tadeusz Jaroszczyk, and Bary Verdegan. 1989. "Temporal Changes in Oil and Air Filter Performance Due to Dust Deposition." *Fluid/Particle Separation J.* 2: 202–207.

Sueoka, Akinori, and Paul S. Malchesky. 1983. "Particle Filtration for Determination of Pore Size Characteristics of Microporous Membranes: Applicability to Plasma Separation Membranes." *Separation Science and Technology* 18: 571–584.

Talcott, Robert. M. 1980. "Selecting and Sizing Door-Oliver Filters. Filtration Leaf Test Procedures. Steel Polymer Filters." MS.

Tanny, G. B.; D. K. Strong; W. G. Presswood, and T. H. Meltzer. 1979. "Adsorptive Retention of *Pseudomonas diminuta* by Membrane Filters." *J. Parenteral Drug Assoc.* 33: 40–51.

Tiller, Frank M. 1955. "The Role of Porosity in Filtration, Part 2, Analytical Equations for Constant Rate Filtration." *Chem. Eng. Progress,* June, 282–290.

_____ . 1975. "What the Filterman Should Know About Theory." *Filtration & Separation,* JulyAug., 386–410.

Tiller, Frank M.; Antoine Alciatore, and Mompei Shirato. 1977. In Chapter 5 of *Filtration Principles and Practice, Part I,* Clyde Orr (Ed.) New York: Marcel Dekker.

Tiller, Frank M. 1990a. "Tutorial: Interpretation of Filtration Data, I." *Fluid/Particle Separation J.* 3: 85–94.

_____, 1990b, "Tutorial: Interpretation of Filtration Data, II, Pilot Plant Filtration of Cottonseed Oil." *Fluid/Particle Separation J.* 3: 157–164.

_____. 1991. "From Pipe to Porous Media Parameters." *Fluid/Particle Separation J.* 4: 1–8.

Treybal, Robert E. 1980. *Mass-Transfer Operations.* New York: McGraw-Hill.

Trottier, R. A., and R. C. Brown. 1990. "The Effect of Aerosol Charge and Filter Charge on the Filtration of Submicrometer Aerosols." Bulletin ITI 0921, TSI, Inc., St. Paul, Minn.

Verdegan, Barry M.; James Draxler, and Harold Fenrick. 1985. "Field Dependence of Particle Electrophoretic Mobility on Non-Polar Liquids." *Particulate Science & Technology* 3: 115– 126.

Verdegan, Barry M. 1986. "Cross-Flow Electrofiltration of Petroleum Oils," *Separation Science & Technology* 21: 603–623.

_____. 1987. "The Effect of Oil Additives on Fuel Water Coalescer Performance," Paper presented at Filttech Conference, Utrecht, Netherlands.

_____. 1989. "Optical Particle Counter Calibration for Nonpolar Applications." Paper presented at Filtech '89, Karlsruhe, Germany.

Verdegan, Barry M. ; Tadeusz Jaroszczyk, and Jeffrey A. Stinson. 1987. "Interpretation of Filter Ratings for Lubrication Systems." Paper presented at 42nd Annual Meeting of the Amer. Soc. Lubrication Engineers, Anaheim, California.

Verdegan, Barry M., and R. A. Jaisinghani. 1980. "Field Study of Ship Bilge Water and Its Implications for Separator Evaluation." *Marine Technology,* Oct., 76–80.

Verdegan, Barry M.; Kendall McBroom; Brian W. Schwandt; Christopher E. Holm, and Lawrence Liebmann. 1992. "Advances in Oil Filter Test Methods." Paper presented at the International Fluid Power Exposition.

Verdegan, Barry M.; Kendall McBroom, and Lawrence Liebman. 1992. "Recent Developments in Oil Filter Test Methods." *Filtration & Separation,* July/Aug.

Verdegan, Barry M.; Brian W. Schwandt, and Christopher E. Holm. 1993. "Development of a Traceable Particle Counter Calibration Standard." Paper presented at the 6th World Filtration Congress, Nagoya, Japan.

Verdegan, Barry M.; Jeffrey A. Stinson, and Lara Thibodeau. 1988. "Accurate Methods for Particle Counting." Paper presented at the 43rd National Conference on Fluid Power, Chicago.

Verdegan, Barry M., and Laura Thibodeau. 1989. "Particle Counting Oil and Water Emulsions." *Particulate Science and Technology* 7: 23–34.

Verdegan, Barry M.; Laura Thibodeau, and Stephen L. Fallon. "Lubricating Oil Condition Monitoring Through Particle Size Analysis." Society of Automotive Engineers Technical Paper 881824. Warrendale, Pa.

Walton, Harris G. 1978. "Laboratory Procedure and Filter for Diatomite Filtration Tests," *Filtration & Separation,* Jan./Feb.

_____. 1981. "Diatomite Filtration—Optimising the Body Feed." *Filtration & Separation.*

Wepfer, R., and J. Schier. 1986. "Production Control of ULPA Filters," Paper presented at the 8th International Symposium on Contamination Control, Milan.

Wepfer, R. 1989. "Quality Tests for ULPA Filters." Paper presented at the 20th Annual Meeting of the Fine Particle Society, Boston.

_____. 1990. "Characterization of ULPA Filters." Paper presented at the International Symposium on Contamination Control, Zurich.

Werynaski, A.; P. S. Malchesky; A. Sucoka; Y. Asanuma; J. W. Smith; K. Kayashima; E. Herpy; H. Sato, and Y Nos. 1981. "Membrane Plasma Separation Toward Improved Clinical Operation." *Transactions of the American Society of Artifical Internal Organs.* XXVII: 539–543.

Williams, C. J. 1992. "Testing the Performance of Spool-Wound Cartridge Filters." *Filtration & Separation,* Mar./Apr., 162–168.

Willis, Max S., and Ismail Tosun. 1980. "A Rigorous Cake Filtration Theory." *Chemical Engineering Science* 35: 2427–2438.

SOME MEASUREMENT-CONVERSION FACTORS

Pressure:

N/m^2 = Pa
1 lbf/in.^{2}, psi, = 6.895 kPa = 144 lbf/ft^2
1 mm Hg = 133 Pa = 1 torr
1 cm Hg = 1.333 kPa ; 1 in. water = 249 Pa
1 atm = $1.013 \cdot 10^5$ Pa = $1.013 \cdot 10^6$ Baryes, or dyn/cm^2
1 bar = 10^6 Baryes = 0.9869 atm

Viscosity: 1 centipoise, cP, = 10^{-3} N · s/m^2 = 10^{-3} Pa · s
= 10^{-3} kg/m · s = 2.89 lbf · sec/ft^2

$\boldsymbol{\nu = \eta/\rho = 0.22t - 180/t}$, where ν = kinematic viscosity, centistokes,
η = absolute viscosity, centipoise,
ρ = density, Specific gravity,
t = Saybolt Universal, seconds.

18°C air: $1.83 \cdot 10^{-5}$ Pa · s

Density: 1 g/cc = 1 Sp.gr. = 62.43 lbm/ft^3 = 8.34 lbm/US gal
= 1 kg/liter = 10^3 kg/m^3

°Bé = 145 – (145/Sp.gr.), *for Sp.gr. greater than 1.0.*
= (140/Sp.gr.) – 130, *for Sp.gr. of 1.0 and less.*

°API = (141.5/Sp.gr.) – 131.5; °Tw = 200 (Sp.gr. – 1)

Surface Tension: 1 dyn/cm = 10^{-3} N/m

Volumetric Flowrate: 1 US gal/min = $6.31 \cdot 10^{-5}$ m^3/s
1 L/min = $1.67 \cdot 10^{-5}$ m^3/s; 1 ft^3/min = $4.72 \cdot 10^{-5}$ m^3/s

---

Area:

1 in.^{2} = $6.45 \cdot 10^{-4}$ m^2; 1 ft^2 = $9.29 \cdot 10^{-2}$ m^2 ; 1 cm^2 = 10^{-4} m^2

INDEX